COMPUTERS
IN
POLYMER SCIENCES

COMPUTERS IN
CHEMISTRY AND INSTRUMENTATION

edited by

James S. Mattson   Harry B. Mark, Jr.   Hubert C. MacDonald, Jr.

COMPUTERS IN POLYMER SCIENCES

EDITED BY

James S. Mattson

*National Oceanic and
Atmospheric Administration
Environmental Data Service
Center for Experiment Design
and Data Analysis
Washington, D.C.*

Harry B. Mark, Jr.

*Department of Chemistry
University of Cincinnati
Cincinnati, Ohio*

Hubert C. MacDonald, Jr.

*Koppers Company, Inc.
Research Department
Monroeville, Pennsylvania*

MARCEL DEKKER, INC.   New York and Basel

Library of Congress Cataloging in Publication Data
Main entry under title:

Computers in polymer sciences.

   (Computers in chemistry and instrumentation ; v. 6)
   Includes bibliographies and indexes.
   1. Polymers and polymerization--Data processing.
I. Mattson, James S. II. Mark, Harry B. III. Mac-
Donald, Hubert C.
QD281.P6C65   547'.84'0285   75-40603
ISBN 0-8247-6368-8

MARCEL DEKKER, INC.
270 Madison Avenue, New York, New York 10016

Current printing (last digit):
10 9 8 7 6 5 4 3 2 1

PRINTED IN THE UNITED STATES OF AMERICA

INTRODUCTION TO THE SERIES

In the past decade, computer technology and design (both analog and digital) and the development of low cost linear and digital
"integrated circuitry" have advanced at an almost unbelievable rate.
Thus, computers and quantitative electronic circuitry are now readily available to chemists, physicists, and other scientific groups
interested in instrument design. To quote a recent statement of a
colleague, "the computer and integrated circuitry are revolution-
izing measurement and instrumentation in science." In general, the
chemist is just beginning to realize and understand the potential
of computer applications to chemical research and quantitative
measurement. The basic applications are in the areas of data acquisition and reduction, simulation, and instrumentation (on-line
data processing and experimental control in and/or optimization in
real time).

At present, a serious time lag exists between the development
of electronic computer technology and the practice or application
in the physical sciences. Thus, this series aims to bridge this
communication gap by presenting comprehensive and instructive
chapters on various aspects of the field written by outstanding
researchers. By this means, the experience and expertise of these
scientists are made available for study and discussion.

It is intended that these volumes will contain articles covering a wide variety of topics written for the nonspecialist but still
retaining a scholarly level of treatment. As the series was con-
ceived it was hoped that each volume (with the exception of Volume 1

which is an introductory discussion of basic principles and applications) would be devoted to one subject; for example, electrochemistry, spectroscopy, on-line analytical service systems. This format will be followed wherever possible. It soon became evident, however, that to delay publication of completed manuscripts while waiting to obtain a volume dealing with a single subject would be unfair to not only the authors but, more important, the intended audience. Thus, priority has been given to speed of publication lest the material become dated while awaiting publication. Therefore, some volumes will contain mixed topics.

The editors have also decided that submitted as well as the usual invited contributions will be published in the series. Thus, scientists who have recent developments and advances of potential interest should submit detailed outlines of their proposed contribution to one of the editors for consideration concerning suitability for publication. The articles should be imaginative, critical, and comprehensive, survey topics in the field and/or other fields, and which are written on a high level, that is, satisfying to specialists and nonspecialists alike. Parts of programs can be used in the text to illustrate special procedures and concepts, but in general, we do not plan to reproduce complete programs themselves, as much of this material is either routine or represents the particular personality of either the author or his computer.

The Editors

PREFACE

Computers and computer technology have found many practical applications in the advancement of polymer science. Indeed, because of the economic importance of polymers to modern society, there has been ample support of research efforts to improve the quality of these products. Much of this research has only been possible because the workers have used computers for calculations and control of experiments. Volume 6 of this series presents results of people working in this area.

The first six chapters are devoted to theoretical considerations of polymerization reactions and to understanding the reaction variables and their influence on polymer products. The final four chapters show ways in which the polymer scientist uses the computer to control his experiments and to simulate data obtained from experimentation on polymeric material.

E. Klesper and A. Johnsen discuss the use of Monte Carlo methods in the simulation of copolymerization reactions. They show the relationship between theoretical models and the results which are actually obtained. Harwood, Kodaira, and Newman have presented methods for calculating compositions of polymers prepared in high conversions. O'Driscoll discusses a computer study of polymerization reactions at the molecular level. In such a study, novel aspects of complex polymerization reaction mechanisms can be determined through simulation leading to a deeper understanding of the reaction scheme.

Rudin shows the merit of machine computation of reactivity

vi

ratios and the variations possible when multicomponent copolymerization data are solved directly for reactivity ratios.

Pittman and Rounsefell present a detailed description of the
chemistry and methods employed in their computer studies of the
Penultimate and Charge-Transfer Polymerization Models. These programs are not included in this volume because they are too long.
Their second chapter presents the method of classifying the reactivity of vinyl monomers in terms of resonance stabilization (Q)
and vinyl group polarity (e). They show a rapid and realistic
evaluation of the Q-e scheme for a monomer which has been copoly-
merized with a series of comonomers.

Niu and Krieger describe the design and implementation of a
rotational rheometer. The rheometer operates under control of a
minicomputer. The minicomputer also acquires and processes the
data for various experiments. Preliminary test results of this
instrument are reported.

An extensive chapter was prepared by Wen and Dole concerning
the computer interfacing of a differential scanning calorimeter
and a thermogravimeter. The authors show the steps in the generation of the software for this task and how this philosophy can be
carried over into other areas. The results are a study of real
time control and data acquisition.

The last two chapters treat two useful data reduction routines. Vescelius presents a method of resolving absorption spectra
or similar curves into its Gaussian and Lorentzian components using
digital computation methods. Although this technique has been done
by an analog computer, digital computers are more available, giving
this method a particular advantage. Misra and Stein give a detailed description of the computer calculation of light scattering from
crystalline polymers. Two approaches, the model approach and the
correlation function approach, are given. Because the developed
theories cannot be solved analytically, evaluation by computer simulation is a necessity.

James S. Mattson
Harry B. Mark, Jr.
Hubert C. MacDonald, Jr.

Monroeville, Pennsylvania

CONTRIBUTORS TO THIS VOLUME

MALCOLM DOLE, Department of Chemistry, Baylor University, Waco, Texas

H. JAMES HARWOOD, Institute of Polymer Science, The University of Akron, Akron, Ohio

A. Ø. JOHNSEN, Institut fur Makromolekulare Chemie, Universitat Freiburg, Freiburg, Germany

E. KLESPER, Institut fur Makromolekulare Chemie, Universitat Freiburg, Freiburg, Germany

YASUTO KODAIRA, Institute of Polymer Science, The University of Akron, Akron, Ohio

IRVIN M. KRIEGER, Department of Macromolecular Science, Case Western Reserve University, Cleveland, Ohio

ASHOK MISRA, Research Department, Monsanto Polymers and Petro-chemicals Co., Springfield, Massachusetts

DANIEL L. NEWMAN, Institute of Polymer Science, The University of Akron, Akron, Ohio

TYAN-FAUNG NIU,* Department of Macromolecular Science, Case Western Reserve University, Cleveland, Ohio

KENNETH F. O'DRISCOLL, Department of Chemical Engineering, University of Waterloo, Waterloo, Ontario, Canada

CHARLES U. PITTMAN, Department of Chemistry, University of Alabama, University, Alabama

THANE D'ARCY ROUNSEFELL, Department of Chemistry, University of Alabama, University, Alabama

ALFRED RUDIN, Department of Chemistry and Chemical Engineering, University of Waterloo, Waterloo, Ontario, Canada

---

*Current affiliation, Photo and Repro Division, GAF Corporation, Binghamton, New York.

RICHARD S. STEIN, Polymer Research Institute, University of
  Massachusetts

LEE E. VESCELIUS, The Firestone Tire and Rubber Company, Central
  Research Laboratories, Akron, Ohio

WALTER Y. WEN, Plastics Laboratory, Honeywell Inc., Hopkins,
  Minnesota

CONTENTS

COMPUTERS
IN
POLYMER SCIENCES

Chapter 1

COMPUTER STUDIES OF REACTIONS
ON SYNTHETIC POLYMERS

E. Klesper

A. Ø. Johnsen

Institut für Makromolekulare Chemie
Universität Freiburg
Germany

I. INTRODUCTION

It has been only in the last few years that greater interest
has arisen in computer Monte Carlo experiments [1-5] and computer
calculations [6-14] of reactions on synthetic polymers. In this

chapter a few typical examples of the use of computers and the results in studying such reactions will be discussed in some detail.

The interest in the Monte Carlo experiments originated from
the finding that reactions on polymers are susceptible to a variety
of influences which are difficult to isolate and to control in
chemical experiments [15,16], and from the difficulties experienced
in obtaining exact and tractable analytical solutions for the simultaneous differential equations describing the different types of
reactions on polymers [17-29]. The kinetic influences on the reaction of polymers may be subdivided:

1. The neighboring group effect
2. The remote group effect
3. The environmental effect

The neighboring group effect accounts for the kinetic influence of
immediately adjacent groups, e.g., monomer units, on a group par-
taking in the chemical reaction. It depends on the nature of the
influencing group: chemical composition, configuration, conformation, and solvation state. The neighboring group effect can be
conveniently investigated by studying the conversion and the sequences of the copolymers which are formed during the reaction on
the polymer. The kinetic influence of neighboring monomer units,
specifically with respect to their chemical composition and their
configuration relative to the reacting group, has been repeatedly
demonstrated for actual reactions on polymers [30-46]. It is to
be expected that conformation and solvation state also have a significant influence in many cases, although there are few detailed
experimental studies as yet [47,48].

The remote group effect pertains to the kinetic influence of
groups which are not immediately adjacent to the reacting group but
which are still present on the same chain. The presence of the
remote group effect has been demonstrated in cases where the influencing group acts as a catalyst for the reacting group [11-13].
Attention was focused thereby on the length and the conformation of
the polymer backbone intervening between influencing group and

reacting group, the distance and the conformation governing the
probability of contact between the two types of groups. Thereby a
calculative treatment was set forth primarily, instead of a Monte
Carlo simulation.

The neighboring and the remote group effect may be termed cooperative effects since two or more groups cooperate with respect
to the reaction in question. Due to the overlapping of different
cooperative regions on the same polymer chain, the rates of reaction of groups are by no means simple and may require for their
elucidation Monte Carlo experiments. Such experiments have the advantage of allowing one to fully define the cooperative influences
governing the reactions, as opposed to actual reaction on polymers,
where the cooperative influences and the cooperative regions are
not known a priori and can only be deduced from the results of the
experiment.

The environmental effect is, in principle, more varied than
the cooperative effect since it is linked to the properties of the
environment of a polymer chain as a whole. Thus the change in concentration and diffusivity of an attacking reagent within a coil of
a polymer in solution, relative to the reagent in the bulk solution,
may be looked upon as an environmental effect. However, a further
change in concentration in the immediate vicinity of a cooperative
region would be looked upon as cooperative if it is caused by that
region. Likewise, the nonaccessibility of a densely coild block
of like monomer units for the attacking reagent in a block copolymer
could be classified as an environmental effect or the nonaccessibility of part of the polymer molecules because of a concentration
gradient for the reagent in the bulk [49]. The effect of the concentration of the polymer and the effect of their length could be
classified as an environmental effect to the extent that the effect
goes beyond that which is due to the change in the state of the
cooperative region connected with the change in concentration and
chain length.

From the foregoing it may be anticipated that Monte Carlo
studies by computers will allow one to fruitfully investigate the

different kinetic influences which are conceivable, particularly the influences of the cooperative type. The results may then be compared with actual reactions on polymers to elucidate the mechanisms of the latter. In the following, several examples from the literature of a more general nature, in which the neighboring group effect influences the conversion and the statistics of the sequences of monomer units, will be covered in depth. It is this type of cooperative kinetics which has been thought to be of importance in most actual reactions on polymers studied so far.

Reactions of polymers for which exact, even if complicted analytical solutions for the kinetics of reaction and the statistics of the polymers can be given should be preferably treated by computer calculations instead of Monte Carlo simulations. This usually results in a reduction of computer time due to the absence of statistical variations inherent in Monte Carlo experiments. An example, which is treated in Sec. III, is the chemical transformation of an atactic copolymer to a homopolymer. Here, the attention is focused on the sequencing of placements in the final homopolymer as determined by the sequencing of the monomer units in the starting copolymer. Both types of sequences can be calculated from relatively few kinetic parameters, whereby the kinetics and the cooperativity of the transformation and the statistics of the intermediary products can be disregarded.

II.   MONTE CARLO COMPUTER EXPERIMENTS

The Monte Carlo experiments carried out to date have proceeded with mainly two aims. First, to investigate the heterogeneity of conversion between chains of copolymers obtained by chemical reaction on homopolymers of relatively low molecular weight. The molecular weight was assumed to be low in the Monte Carlo experiments because only then does the heterogeneity of conversion become large. The second aim was to study the conversion and the sequencing of monomer units in copolymers derived from homopolymers of relative high molecular weight. Here the conversion and the probabilities

of the sequences in the simulated copolymer chains do not vary
greatly between chains, i.e., the heterogeneity of conversion and
of probabilities of sequences is low.

For both types of studies it was supposed that there are only
two states for a given monomer unit in the chain, an unreacted state
and a reacted state. Furthermore, it was assumed that there exists
only a kinetic influence of the next neighboring monomer unit, i.e.,
whether reacted or not, but not of monomer units further removed
along the chain, nor intermediates occurring during the conversion
of a given monomer unit. It was supposed, furthermore, that the
chains were to be accessible to the attacking reagent to the same
extent for different chains as well as for different parts of the
same chain and that the concentration of the reagent did not change
during the reaction. The latter supposition is, of course, equivalent to assuming a reaction which is quasizeroth order with respect to reagent. Also, provision was made in the Monte Carlo
experiments for connecting the chains end to end to avoid monomer
units on the ends with only one neighboring monomer unit.

It should be noted that the basic assumptions of the simulation
of reactions on polymers also allows application of the result to
other processes taking place on a homopolymer, e.g., adsorption of
a low molecular weight species, as long as there are only two states
which have a different kinetic influence as next neighbors to the
reacting site. Moreover, the Monte Carlo procedures can be easily
changed to accommodate further assumptions about the reactions,
the assumption of more than two states, the kinetic influence of
four neighboring monomer units, or starting the simulated reaction
on a copolymer instead of a homopolymer.

A. Heterogeneity of Conversion for Short Chain Copolymers Derived by Reactions on Homopolymers

The work to be described in this section has been carried out
by Litmanovich et al. [1]. The assumed homopolymers were of relatively low molecular weight, in which, for some simulation runs,

the cyclized chains were assumed to be monodispersed with respect to
degree of polymerization and polydispersed for others.

1. Computer Program

A computer M-20 was used to process cyclized chains of 10, 20, 30,
40, 50, 60, 75, 100, and 200 monomer units. An unreacted monomer
unit was represented by 0 and a reacted unit by 1 in the computer
memory. At the start of the simulation, all monomer units were in
the unreacted state. Because only the kinetic influence of the two
next neighbors was considered, three reaction rate constants $k(100)$,
$k(001^+)$, and $k(101)$ applied, where $k(001^+)$, for instance, is the
rate constant for the reaction of a central 0 unit flanked by one 0
and one 1 unit, whereby the + sign indicates that the order may be
either 001 or 100. It could be shown [2] that the statistics of the
sequencing of monomer units in copolymers at a given conversion depends only on the ratio $k(000):k(001^+):k(101)$ and not on the absolute magnitude of the $k(0$. Thus only that ration had to be preselected for a given run, in principle.

The calling of the monomer units to test them for reaction was
carried out by proceeding from one monomer unit to the next neighboring monomer unit, starting from a given unit of the chain and
going around the macrocycle returning to the starting point. The
calling of the monomer units provided a useful simulated reaction
time by counting the number of completed passes around the macrocycle. Each monomer unit 0 which was called during a pass was
tested for reaction $0 \rightarrow 1$ by determining the state of the two neighboring monomer units and generating a pseudorandom number. If the
pseudorandom number did not exceed a preselected value determined by
the ratio of the reaction rate constants and the pseudorandom number
range, the conversion was supposed to take place; otherwise, the
monomer unit was considered to remain unreacted. However, even if
the test for conversion of a monomer unit was positive, the reaction $0 \rightarrow 1$ was not effected immediately, but only after a completed
pass. After each pass the conversion of the cyclized chain was
noted and the passes continued until a total conversion of 90% of

the monomer units was reached. This process was repeated several dozen times with a cyclized homopolymer chain to obtain data for the heterogeneity of conversion between chains after a given number of passes.

The calling of the monomer units by stepping unidirectionally from one monomer unit to the next along the cyclized chain does not correspond to the desired type of actual reactions on polymers. During actual reactions, the chances of conversion for monomer units 0 surrounded, for example, by 0 and 1, is the same at a given time for all such monomer units along the chain, i.e., the conversion occurs randomly for a monomer unit in a given surrounding and not unidirectionally. Thus the calling of the monomer units should be of a random nature. Moreover, in the program described, the postponing of the reaction 0 → 1 to the time of the completion of a pass is related to the assumption that there are a relatively large number of simultaneous reactions of neighboring monomer units. This, however, does also not correspond to reality because most reactions on polymers possess half times of conversion which range from minutes to many hours. These two features of the program may lead to distortions with respect to the time dependent conversion and the statistics of sequences of the simulated copolymers. However, as is shown in Sec. II.B, the distortions may be suppressed by decreasing the preselected balues with which the pseudorandom numbers are to be compared, without changing the ratio of rate constants, thereby reducing the number of individual reactions 0 → 1 per pass. It may be noted at this point that in no case do distortions arise on account of simultaneous reactions of neighboring monomer units if the ratio of rate constants $k(000):k(001^+):k(101) = 1:1:1$, since then it is of no consequence whether or not a neighboring monomer unit has reacted. The distortion caused by the unidirectional nture of the calling will, however, still prevail with this ratio of rate constants, provided each called 0 unit is reacted.

2. Results of Monte Carlo Experiments

Two ratios of rate constants were studied, the ratio $k(000):k(001^+):$
$k(101) = 1:1:1$ and the ratio $k(000):k(001^+):k(101) = 1:5:100$. With
the first ratio there is no difference in the kinetic influence of
the two types of monomer units, and the sequencing of the monomer
units in the resulting copolymers must therefore be random if the
simulation procedure is correct. With the second ratio 1:5:100, the
reaction is autoaccelerating with increasing conversion because
after the reacted units are formed in the chain, the two higher rate
constants come into effect. The sequencing of the monomer units
must possess block character since a once-formed reacted unit leads
to an increased rate of reaction for the unreacted units present as
a next neighbor of a reacted unit. Actual reactions on polymers
with a ratio of rate constants leading to randomness, i.e., Bernoullian statistics, have been found in a number of cases [49-53], as
well as some reactions with ratios of rate constants leading to
autoacceleration and a sequencing of block character [36,42,54].

The results of the Monte Carlo experiments for the ratio 1:1:1
are shown in Fig. 1. The heterogeneity of conversion is presented
as the weight fraction $w[P(B)]$ at a given conversion $P(B)$, where
$P(B)$ is the estimate of the probability of finding a reacted monomer
unit in the chain. $P(B)$ is only an estimate of the probability because the number of runs employed for obtaining the data is limited.
The degree of polymerization for the simulated, cyclized chains is
30 (Fig. 1, plot I), 50 (plot II), and 200 (plot III). Four weight
functions $w[P(B)]$ are shown in each plot with the mean conversion
$\overline{P(B)} = 0.2$ (a), 0.3 (b), 0.5 (c), and 0.7 (d). Overall comparison
of plots I to III shows that the heterogeneity of conversion is
largest for the lowest degree of polymerization and decreases with
increasing length of the chain, as one would expect. Comparing
the heterogeneity of conversion within a given plot shows that
apparently the heterogeneity is largest somewhere in the middle
range of conversion.

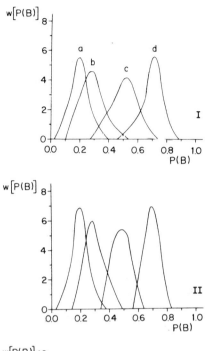

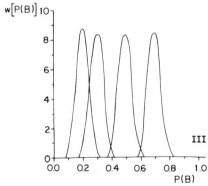

FIG. 1. Heterogeneity of conversion for the ratio of rate constants 1:1:1 plotted as weight function w[P(B)] versus P(B). The function w[P(B)] represents the fraction of simulated copolymers of a given conversion P(B), where P(B) is the probability of finding a reacted monomer unit in the chain. The w[P(B)] are being obtained by repeated simulation of reaction on cyclized chains of 30 (I), 50 (II), and 200 (III) monomer units. Four weight functions each are shown for the mean conversion $\overline{P(B)}$ = 0.2 (a), 0.3 (b), 0.5 (c), and 0.7 (d).

In Fig. 2 the results are presented for the ratio of rate constants $k(000):k(001^{+}):k(101) = 1:5:100$ in the same way as in Fig. 1.
This autoaccelerating ratio results in an overall larger heterogeneity
of conversion than does the random ratio. This is not surprising
because once a given run has reached, on account of statistical variation, a conversion at a given number of passes which is higher
than the average conversion, then this difference of conversion
tends to increase at progressing conversion with the autoaccelerating ratio more than with the random ratio. Again, with increasing
degree of polymerization the heterogeneity decreases, as seen from
the comparison of plots I to III of Fig. 2. Within a given plot,
however, the maximum of the heterogeneity seems to be moved to higher conversion when compared to the random ratio of rate constants.

Similar results as in Figs. 1 and 2 are obtained if, rather
than monodispersed, cyclized chains, one assumes for the simulation
polydispersed cyclized chains possessing a so-called most probable
distribution of the degree of polymerization P_n:

$$w[P_n] = \frac{P_n}{(\overline{P_n})^2} \exp \left(- \frac{P_n}{\overline{P_n}} \right) \tag{1}$$

where $w[P_n]$ is the weight fraction of a given degree of polymerization and $\overline{P_n}$ is the average degree of polymerization. In Fig. 3
results of the Monte Carlo experiments for $\overline{P_n} = 50$ and the two
ratios 1:1:1 (plot I) and 1:5:100 (plot II) are seen. The heterogeneity of conversion for the ratio 1:1:1 appears to be larger for
the polydispersed chains (Fig. 3, plot I) than for the monodispersed chains (Fig. 1, plot II). For both ratios the heterogeneity
for the polydispersed chains passes through a maximum during the
course of the reaction.

One should expect that with chains of low degree of polymerization not only the conversion $P(B)$ is heterogeneous but also the
probabilities of sequences of monomer units. This kind of heterogeneity has been briefly investigated, as discussed in Sec. II.B.

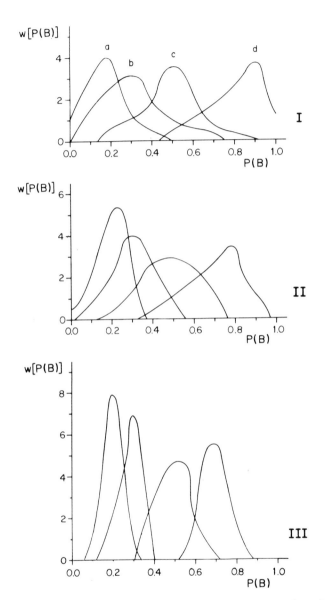

FIG. 2. Heterogeneity of conversion for the ratio of rate constants 1:5:100. (Meaning of symbols same as for Fig. 1.)

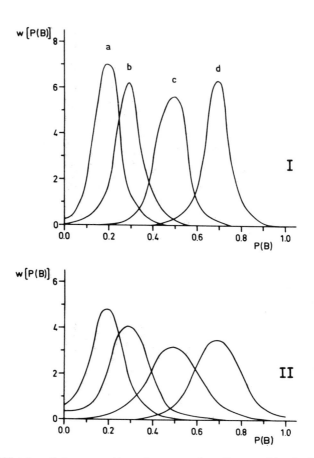

FIG. 3. Heterogeneity of conversion for cyclized chains with
normal distribution of degree of polymerization. Average degree of
polymerization $\overline{P_n}$ = 50 for the ratios of rate constants 1:1:1 (I)
and 1:5:100 (II). Four weight functions w[P(B)] are shown for the
average conversion $\overline{P(B)}$ = 0.2 (a), 0.3 (b), 0.5 (c), and 0.7 (d).

B. Sequences of Monomer Units in Copolymers
Derived by Reactions on Homopolymers

The work in this section has been reported by Klesper et al.
[2]. The basic features of the model were the same as previously
described, e.g., a cooperative chemical reaction was assumed with
the monomer units being in either an unreacted or a reacted state
and with a kinetic influence of only next neighboring monomer units.
The cyclized polymers were, however, only monodispersed with a
degree of polymerization of P_n = 1000.

1. Computer Programs
An IBM 7040 digital computer with Fortran IV programming was employed. The pseudorandom numbers necessary for the program were
generated with the function RDM (Share 1360). The calling of the
monomer units before testing them for reaction was carried out in
two ways: either nonrandomly, progressing in one direction from
one monomer unit to the next neighboring monomer unit as in Sec.
II.A, or else randomly by means of a pseudorandom number. After
calling a particular monomer unit, it was determined whether this
unit was 0. If so, the state of the two next neighboring monomer
units was determined. A pseudorandom number was generated, whereby
the random number range was 0 to 1. If the random number fell into
the range 0 to p( ), the called monomer unit was considered to have
reacted and the conversion 0 → 1 was effected immediately in the
memory of the computer. The p( ) was thereby the reaction probability [0 < p( ) < 1] and selected from p(000), p(001$^+$), and p(101)
depending on which neighbors the called monomer units possessed.
Since p(000):p(001$^+$):p(101) = k(000):k(001$^+$):k(101), the substitution of the ratio of reaction probabilities for the ratio of reaction rate constants had no effect on the statistics of the sequencing of the monomer units at a given conversion. After sufficiently large increments on the simulated time scale, the number of
0 and 1 units and the number of specific sequences of such units,
i.e., specific dyads, triads, tetrads, and pentads of monomer units,
were counted and the results printed. In case several repeated runs

were averaged, only the printed results at the same simulated time
of the reaction were taken for the averaging, of course.

With the nonrandom call procedure, the reaction $0 \to 1$ was not
postponed until the completion of a pass around the cyclized chain,
as opposed to the computer program of Sec. II.A. Thus the quasi-
simultaneous reactions of neighboring monomer units during a given
pass have been eliminated. On the other hand, a directionally
biased growth of blocks of reacted monomer units and the quasi-
simultaneous formation of new blocks of reacted units during a given
pass are still inherent in this nonrandom call procedure. The distortion caused in the conversion and in the statistics of sequences
of the copolymers could be eliminated, however, by reducing the
magnitude of the reaction probabilities by a common factor without
changing their ratio. Then fewer reactions $0 \to 1$ occur during a
pass with the consequence that directional bias and simultaneous
growth of blocks of reacted units is reduced. The nonrandom call
program may then give the same statistics at a given conversion for
the copolymers as the more realistic random call procedure, as may
be derived from Fig. 4.

In Fig. 4 the estimates of the probabilities of the A-centered
triads obtained by the nonrandom call procedure are plotted versus
the estimates of $P(B)$ as points (A = unreacted unit, B = reacted
unit). The filled points ($\bullet$, $\blacksquare$, $\blacktriangle$) have been obtained by the set
of reaction probabilities $p(000)$; $p(001^+)$; $p(101) = 0.1$; 1.0; 0.1,
respectively, and the open points (o, $\square$, $\triangle$) for the set $p(000)$;
$p(001^+)$; $p(101) = 0.001$; 0.01; 0.001. Reducing the first set by
the common factor 0.01 obviously moves the points towards the drawn-
out lines which represent the results for the random call program
with the first set 0.1; 1.0; 0.1. For all results of Fig. 4 the
heterogeneity of conversion and of triad probabilities has been
greatly reduced by averaging the results of five runs with 1000
monomer units each. In contrast to the nonrandom call procedure, a
reduction of the set of reaction probabilities has no effect on the
statistics at a given conversion obtained by the random call procedure (not shown), as to be expected for a correct simulation.

P(triads)

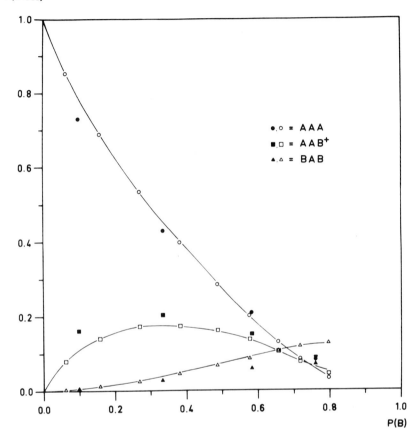

FIG. 4. Comparison between nonrandom call and random call computer programs. Probabilities for A-centered triads obtained by
nonrandom call procedure with the set of reaction probabilities
$p(000)$; $p(001^+)$; $p(101)$ = 0.1; 1.0; 0.1 shown as filled points
($\bullet$, $\blacksquare$, $\blacktriangle$), those of set 0.001; 0.01; 0.001 as open points (o, □,
Δ). Probabilities of A-centered triads obtained by random call
procedure shown as drawn-out lines. For all data shown, five computer runs with 1000 monomer units each were averaged.

With other sets of reaction probabilities a reduction of their magnitude is not necessary for the nonrandom call procedure to yield a
result similar to that obtained with the random call program. This
is demonstrated for the set $p(000)$; $p(001^+)$; $p(101)$ = 0.01; 0.1;

1.0 in Fig. 5. Again, the points apply to the nonrandom call procedure and the drawn-out curves to the random call procedure. Five
runs of 1000 monomer units each have been averaged.

P(triads)

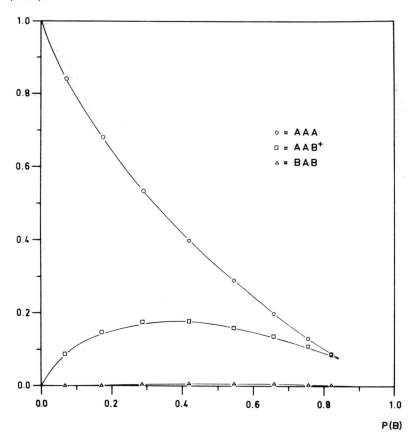

P(B)

FIG. 5. Comparison between nonrandom call and random call
computer programs. Probabilities of A-centered triads obtained by
nonrandom call procedure with the set of reaction probabilities
$p(000)$; $p(001^+)$; $p(101) = 0.01$; 0.1; 0.1 shown as points (o, □, Δ),
those obtained by random call procedure with the same set shown as
drawn-out lines. Five runs of 1000 monomer units each were
averaged.

To test the random call procedure, the probabilities of triads obtained with the random call procedure for the random set 1.0; 1.0; 1.0 were compated with the calculated Bernoullian values, as Fig. 6 shows in part. The probabilities of the A-centered triads obtained

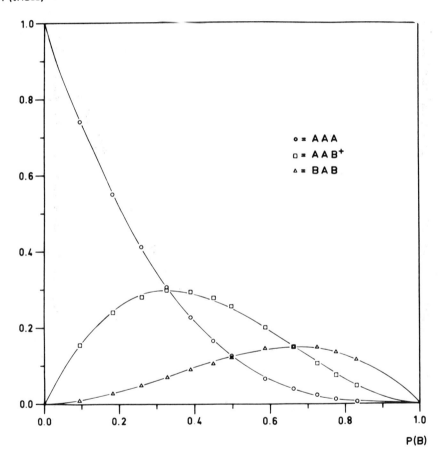

FIG. 6. Probabilities of A-centered triads obtained by random call procedure with the set of reaction probabilities p(000); p(001+); p(101) = 1.0; 1.0; 1.0 shown as points (o, □, Δ), those calculated on the basis of Bernoullian statistics shown as drawn-out lines.

by computer are given as points and agree well with the calculated
probabilities shown as drawn-out curves. In order to reduce the
statistical variation, i.e., heterogeneity of triads and of P(B),
again five runs of 1000 monomer units each were averaged. As
Fig. 6 also shows, this is apparently sufficient to depress the
variation to an acceptable level. In order to obtain more corrob-
oration of the validity of the random call procedure, one should
also test runs for other sets of reaction probabilities, e.g.,
autoaccelerating and autodecelerating sets. However, there appear
to exist no analytical solutions to the general rate equations of
reactions on polymers which have not been derived without significant simplifications.

With the random call procedure, the indexed monomer units of
the cyclized chains of 1000 monomer units were called by the first
three digits of a pseudorandom number. If a reacted unit was called, another random number was generated until an unreacted unit was
found. In order to save computer time and for obtaining a realistic time scale for the simulated reaction, the unused digits of
that last generated random number were utilized to test for conversion $0 \rightarrow 1$. The total count C of the random numbers generated is
then proportional to the reaction time of an actual reaction on a
polymer if the latter reaction is first order with respect to the
reacting triads of monomer units (AAA, AAB$^+$, BAB) and zeroth order
with respect to reagents. To give a point of reference for the
computing times required, a run with 1000 monomer units needs for
one of the slowest sets of reaction probabilities encountered,
p(000); p(001$^+$); p(101) = 1.0; 0.1; 1.01, 55 minutes to reach C =
17200 with a conversion of P(B) ≈ 0.64. Other approximate computing
times can then be derived from Figs. 7 and 8. The computing time can
be reduced by a decreasing table technique which avoids calling by
random number such monomer units which have already reacted [55].

In the following the described random call procedure is used
exclusively.

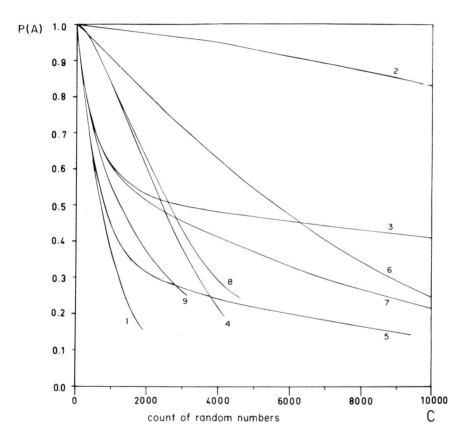

FIG. 7. Probability of monomer unit A P(A), plotted versus coubt of random numbers C for nine different sets of reaction probabilities p(000); p(001+); p(101). Numerals on curves provide key to sets of Table 2.

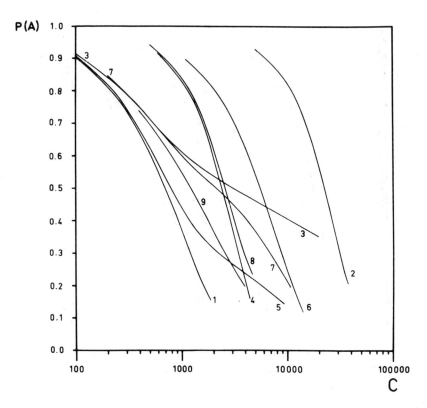

FIG. 8. Plot like in Fig. 7 but with logarithmic scale for C.

2. Heterogeneity of Conversion
    and of Sequences

The heterogeneity of conversion and that of sequences of monomer
units was also investigated by the random call procedure. Taking
the autoaccelerating and autodecelerating sets $p(000)$; $p(001^{+})$;
$p(101)$ = 0.01; 0.1; 1.0 and 1.0; 0.1; 0.01 as examples, the estimates of the standard deviation, $\tilde{\sigma}$, and of the standard error of
the mean, $\tilde{\sigma}_{m}$, are given in Table 1 as a measure of the heterogeneity

TABLE 1

Estimates of the Standard Deviation $\tilde{\sigma}$ and of the
Standard Error of the Mean $\tilde{\sigma}_m$ (in parentheses)

Five Repeat Runs with $P_n = 1000$ Each

p(000); p(001+); p(101)	0.01; 0.1; 1.0				1.0; 0.1; 0.01			
$\overline{P(B)}$	0.171	0.427	0.557	0.746	0.209	0.409	0.522	0.611
C	10000	20000	25000	35000	300	1200	4200	12200
P(B)	12 (3.8)	27 (15)	69 (22)	50 (16)	4.7 (1.5)	10 (3.2)	9.5 (3.0)	6.9 (2.2)
P(AAA)	26 (8.3)	63 (20)	66 (21)	38 (12)	13 (4.0)	10 (3.3)	1.4 (0.45)	– –
P(AAB+)	18 (5.6)	21 (6.6)	17 (5.5)	17 (5.3)	13 (4.0)	21 (6.7)	17 (5.5)	6.0 (1.9)
P(BAB)	2.5 (0.78)	3.2 (1.0)	2.0 (0.63)	2.0 (0.63)	4.7 (1.5)	18 (5.6)	10 (3.3)	8.8 (2.8)
P(BBB)	12 (3.8)	41 (13)	69 (22)	63 (20)	0.69 (0.22)	2.0 (0.63)	7.6 (2.4)	13 (4.1)
P(ABB+)	14 (4.3)	20 (6.2)	20 (6.2)	10 (3.2)	5.4 (1.7)	5.4 (1.7)	13 (4.2)	12 (3.7)
P(ABA)	7.3 (2.3)	6.6 (2.1)	6.0 (1.9)	3.2 (1.0)	5.7 (1.8)	7.8 (2.5)	10 (3.2)	10 (3.2)

$\tilde{\sigma} \cdot 10^3$ $\tilde{\sigma}_m \cdot 10^3$

of conversion and of triads. The data have been obtained with only
five repeat runs of 1000 monomer units each. The average conversion is given as $\overline{P(B)}$ and the count of the random number as C. The
values for $\tilde{\sigma}$ and $\tilde{\sigma}_m$ clearly show that the heterogeneity is strongly
dependent on the set of reaction probabilities as well as on the
probability in question. Some probabilities show a greater heterogeneity with the autoaccelerating set than with the autodecelerating set, while other probabilities show a reverse behavior. Moreover, there does not seem to exist a simple relationship between
the magnitude of the probabilities and their $\tilde{\sigma}$ and $\tilde{\sigma}_m$. For instance, P(AAA) drops with increasing $\overline{P(B)}$ without a corresponding
decrease of $\tilde{\sigma}$ and $\tilde{\sigma}_m$, as seen by comparison of Fig. 13 with Table 1.
In view of the values of $\tilde{\sigma}_m$ of about 0.001 to 0.020 for the five
repeat runs in Table 1 and for runs with other sets of reaction
probabilities one might consider five repeat runs as a suitable
compromise between precision and computing time. All data in the
following discussion are therefore based on five repeat runs of
1000 monomer units each, unless noted otherwise.

3. Kinetics of Conversion

Plotting the count C of random numbers versus log P(B) for the random set p(000); p(001$^+$); p(101) = 1.0; 1.0; 1.0, results in a
straight line, i.e., a kinetic first-order plot. This is to be expected if C is a suitable time scale for the random call procedure.
A first-order relationship between C and log P(B) is, however, not
to be expected for sets with unequal reaction probabilities, even
if the kinetics is first order with respect to the individual reacting triads 000, 001$^+$, and 101. This can be seen from the appropriate second-order rate equation for an actual reaction on a polymer

$$\frac{dP(A)}{dt} = -k(AAA)[R]P(AAA) - k(AAB^+)[R]P(AAB^+)$$

$$-k(BAB)[R]P(BAB) \qquad (2)$$

where [R] is the concentration of the attacking reagent. With unequal k( ) no first-order behavior of P(A) with time, and therefore

also of P(A) with C, can be expected even if [R[= constant. If,
however, p(000); p(001^{+}); p(101) = 1.0; 1.0; 1.0, this corresponds
to k(AAA) = k(AAB^{+}) = k(BAB) = K, and one obtains first-Order behavior of P(A) with time when [R[= constant.

$$\frac{dP(A)}{dt} = -K[R[P(A) = -K'P(A) \qquad K' = \text{constant} \qquad (3)$$

Since no provision is made for changing [R[during the simulation
the first order holds also for P(A) versus C.

For nine different sets of reaction probabilities the simulated behavior of P(A) versus C is shown in Fig. 7. The numerals

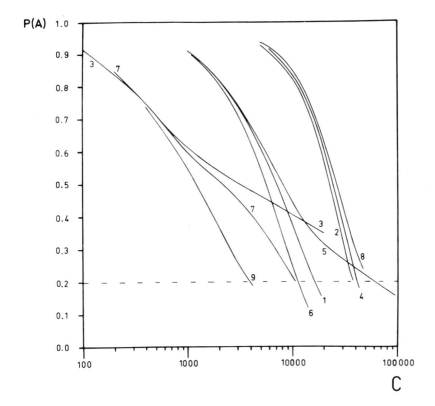

FIG. 9. Plot like in Fig. 7 but with logarithmic scale for C
and with reduction of sets of reaction probabilities by common
factor for adjusting p(001^{+}) to 0.1.

1   0.1; 0.1; 0.1/2 → 0.01; 0.1; 1.0/3 → 1.0; 0.1; 0.01
4   0.01; 0.1; 0.1/5 → 0.1; 0.1; 0.01/6 → 0.1; 0.1; 1.0
7   1.0; 0.1; 0.1/8 → 0.01; 0.1; 0.01/9 → 1.0; 0.1; 1.0

written unto the curves provide the key for the sets given in
Table 2. The sets of Table 2 have been chosen to cover the major
possibilities for varying the ratios of the reaction probabilities.
At least one of the reacting probabilities within a set is put at
1.0 to save computing time. The curves in Fig. 7 indicate that with a
a large $p(000)$, i.e., $p(000) = 1.0$, the rate of the simulated reaction is comparatively high during the first part of conversion. On
the other hand, when $p(101) = 1.0$ the rate is comparatively high
during the last part of conversion. The influence of a large $p(001^+)$
on the rate appears to be more evenly distributed over the middle
range of the reaction. This kind of behavior is based on the presence of a high number of reacting triads AAA in the beginning and a
relatively high number of reacting triads BAB at the end of the reaction, while the AAB^+ triads are present in relatively large numbers over a wider, middle range of conversion.

In Fig. 8 the plot of Fig. 7 is extended to data at higher C
by using a logarithmic scale in order to accommodate the slowly

TABLE 2

Sets of Reaction Probabilities for Simulation
of Reactions on Polymers

Set	$p(000)$	$p(001^+)$	$p(101)$
1	1.0	1.0	1.0
2 ⎤	0.01	0.1	1.0
3 ⎦	1.0	0.1	0.01
4 ⎤	0.1	1.0	1.0
5 ⎦	1.0	1.0	0.1
6 ⎤	0.1	0.1	1.0
7 ⎦	1.0	0.1	0.1
8 ⎤	0.1	1.0	0.1
9 ⎦	1.0	0.1	1.0

Note: Bracketed pairs indicate opposing sets.

reacting sets at higher conversion. The graph is thus useful for comparing the computing time needed for reaching a given, substantial conversion. It is also useful for finding the P(A) versus C behavior, without performing computer runs, when reducing the sets of reaction probabilities by a common factor. As may be immediately clear, such reduction does not change the shape of the corresponding curve in the log C plot since the shape depends on the ratio of reaction probabilities, and only moves the curve in a direction parallel to the logarithmic axis. Thus, if set 1 is reduced by the common factor 0.1, then C at each P(A) of curve 1 is to be multiplied by 10 to obtain the curve of the new 0.1; 0.1; 0.1. This has been utilized in Fig. 9 to shift $p(001^+)$ of all 9 sets to 0.1. All 9 curves of Fig. 8 are now directly comparable with respect to the influence of varying the remaining two reaction probabilities. For instance, the sets 5, 1, and 6 are now 0.1; 0.1; 0.01; 0.1; 0.1; 0.1, and 0.1; 0.1; 1.0, respectively. All three sets are equal, save for p(101) which varies from 0.01 over 0.1 to 1.0. It is therefore not surprising that during the first half of the reaction the corresponding curves are situated close to each other while they progressively spread apart during the last half of the reaction, the rate of conversion being largest for set 6 followed by set 1 and set 5.

4. Kinetics of Dyads and Triads

Plotting the probabilities of dyads and triads versus the parameter of conversion, P(B), results also in information about the influence of different ratios of reaction probabilities on the simulated reaction. The plotting versus P(B) instead of versus C, or log C, has been chosen for reasons to be discussed later. In Figs. 10 to 12 the change of the probabilities of the three possible dyads AA, AB^+, and BB with conversion is shown. Inspecting Fig. 10 with the plot P(AA) versus P(B), it is obvious that the most pronounced differentiation between different sets of reaction probabilities is obtained at P(B) ≈0.5. This observation holds also for the plot P(BB) versus P(B) in Fig. 11, although it may be noted that a given

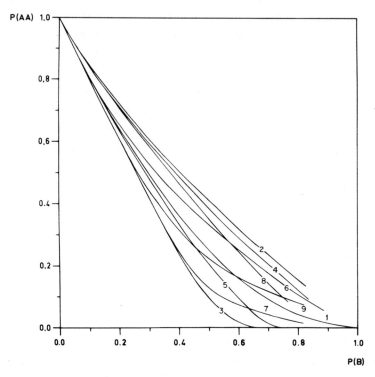

FIG. 10. Probability of dyad AA, P(AA), versus P(B) for nine sets of reaction probabilities. Numerals on curves provide key to sets of Table 2.

curve of Fig. 11 generally cannot be the mirror image of a curve with the same set of reaction probabilities of Fig. 10. The differentiation between different sets of reaction probabilities becomes much stronger for the plot $P(AB^+)$ versus P(B) in Fig. 12, whereby the maximum is again seen at P(B) ≈0.5.

The qualitative behavior of the curves in Figs. 10 to 12 may be rationalized if it is realized that a relatively large value of p(000) should produce a tendency toward alternation of monomer units A and B in the copolymers, while relatively large values for $p(001^+)$ and p(101) should generate a sequencing of monomer units which possesses block character. The term "relatively large" is to mean, thereby, relatively large with respect to the remaining (two or one)

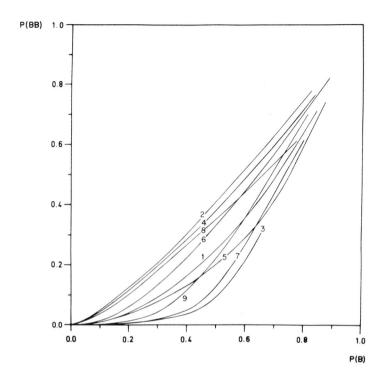

FIG. 11. Probability of dyad BB, P(BB), versus P(B) for nine sets of reaction probabilities. Numerals on curves provide key to sets of Table 2.

reaction probabilities. A relatively large p(000) favors formation of the triad ABA, and consequently of BAB, which are characteristic of an alternating statistics of triads. Relatively large values for p(001$^+$) and p(101) favor the formation of blocks by creating, during reaction, B-units which are preferably adjacent to already existing B-units. Accordingly, the sets of reaction probabilities 2, 4, 6, and 8 should lead to block character, as is confirmed by the excess of AA and BB dyads in Figs. 10 and 11 and a deficiency of AB$^+$ dyads in Fig. 12, whereby the curves 1 of the random set of reaction probabilities serve as the appropriate reference. If the magnitude of the distances from curves 1 are taken as simple measure of the deviation from random dyad statistics, it is seen

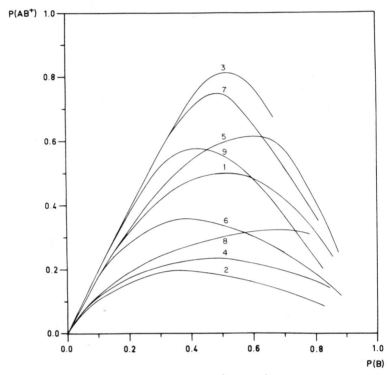

Fig. 12. Probability of dyad AB$^+$, P(AB$^+$), versus P(B) for nine sets of reaction probabilities. Numerals on curves provide key to sets of Table 2.

that the block character decreases for the sets in the order 2, 4, 6, or 8, as one would probably expect on the basis of the change in p(001$^+$) and p(101) going from one set to the next. The crossing of the curves 6 and 8 requires some explanation. The relatively large p(101) for set 6 leads to block character for this set in the last part of the reaction in view of the larger amount of BAB present at this stage. The relatively large p(001$^+$) of set 8 brings about block character at the beginning and at the middle range of conversion, because then the amount of AAB$^+$ to react is larger than at the end of the reaction. This varying block character leads to the crossover of the curves 6 and 8.

Turning to the remaining sets 3, 5, 7, and 9 in Figs. 10 to 12, these create a tendency toward alternation as seen by deficiency of AA and BB and the excess of AB$^+$ when compared to curve 1. Their qualitative behavior may be explained on account of their large p(000) in a similar way as for the sets creating block character.

The dyad probabilities of Figs. 10 to 12 have been plotted versus P(B) instead of C, or log C, because the curves depend then only on the ratio p(000):p(001$^+$):p(101) instead of the absolute magnitude of the p( ). Thus the curves for set 1 → 1.0; 1.0; 1.0 are also valid for the set 0.5; 0.5; 0.5 or any other set obtained by reduction with a common factor. This may, of course, be verified by computer runs, but can also be verified by means of the rate equations for reactions on polymers. The procedure for establishing the rate equations has been discussed [8,56]. The rate equation for the dyad AA is accordingly

$$\frac{dP(AA)}{dt} = -2k(AAA)[R]P(AAA) - k(AAB^+)[R]P(AAB^+) \tag{4}$$

For the computer simulation of reactions on polymers, Eq. (4) may be rewritten

$$\frac{\Delta P(AA)}{\Delta C} = -2p(000)\tau P(AAA) - p(001^+)\tau P(AAB^+) \tag{5}$$

i.e. dt is replaced by ΔC, the concentration [R] by a constant τ, which is computer and program dependent, and the k( ) are replaced by p( ). In a similar way one may rewrite Eq. (2):

$$\frac{\Delta P(A)}{\Delta C} = -p(000)\tau P(AAA) - p(001^+)\tau P(AAB^+) - p(101)\tau P(BAB) \tag{6}$$

Dividing Eq. (5) by Eq. (6) and replacing $\Delta P(A)$ by $-\Delta P(B)$, in view of P(A) + P(B) = 1, leads to

$$\frac{\Delta P(AA)}{\Delta P(B)} = -\frac{2p(000)P(AAA) + p(001^+)P(AAB^+)}{p(000)P(AAA) + p(001^+)P(AAB^+) + p(101)P(BAB)} \tag{7}$$

Noting that each term in the quotient of the right hand side of Eq. (7) contains a reaction probability, only the ratio p(000):p(001$^+$): p(101) may have an influence on the plot of P(AA) versus P(B). Similar reasoning applies the other dyads and also the triads.

The results of the computer experiments for triad probabilities versus P(B) are seen in Figs. 13 to 18. The greatest differentiation between sets of reaction probabilities is found in the middle range of conversion between P(B) = 0.4 and P(B) = 0.7. About the same degree of differentiation is found for the triads AAA (Fig. 13), BBB (Fig. 14), ABA (Fig. 15), BAB (Fig. 16), followed by AAB[+] (Fig. 17) and ABB[+] (Fig. 18). The degree of differentiation between sets of reaction probabilities is of importance in foreseeing the possibilities for varying the statistics of an actual reaction on a polymer by varying the ratio of the reaction rate constants. It is also

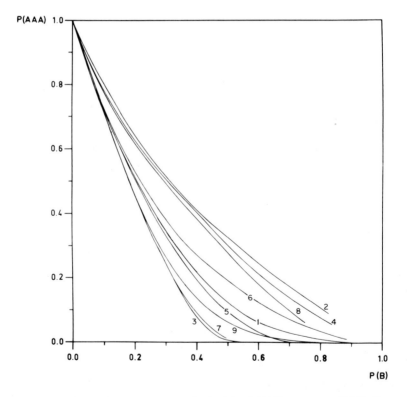

FIG. 13. Probability of triad AAA, P(AAA), versus P(B) for nine sets of reaction probabilities p(000); p(001+); p(101). (See Table 2 for key to sets.)

useful for showing the type of sequence, at a given conversion
range, which is likely to give the most accurate result when determining the ratio of reaction rate constants from sequence probabilities of an actual reaction.

To qualitatively rationalize the shape of the curves of the
triads in dependence of the reaction probabilities is more difficult
than for dyads. This is because the informational content of a triad statistics is larger than can be fully described in terms of
block character or tendency toward alternation. Nevertheless, a
first understanding can be gained in these terms. For this it is

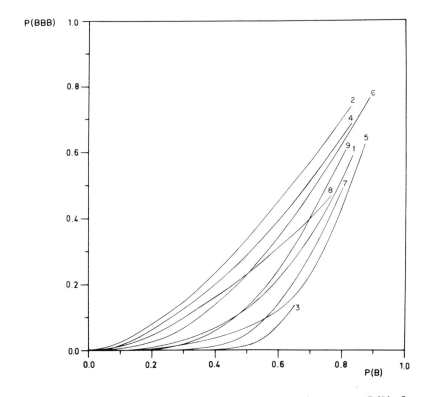

FIG. 14. Probability of triad BBB, P(BBB), versus P(B) for
nine sets of reaction probabilities. (See Table 2 for key to sets.)

convenient to rely on the behavior of the triads ABA and BAB and, secondly, on AAA and BBB. As Figs. 15 and 16 show, the sets of reaction probabilities 2, 4, 6, and 8 generate less ABA and BAB than would be expected for a random statistics (curve 1). Thus a block character is indicated as expected from the relative magnitude of $p(000)$ on one hand and $p(001^+)$, $p(101)$ on the other. The block character is confirmed by the excess of AAA and BBB for these sets in Figs. 13 and 14 when compared to curve 1. It is more difficult to understand the behavior of AAB^+ and ABB^+ for sets 2, 4, 6, and 8 in Figs. 17 and 18. While there is less AAB^+ for sets 2, 4, 8, as

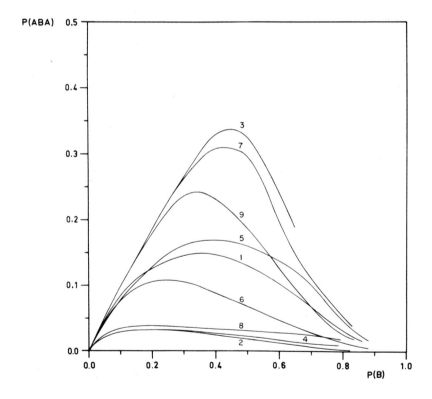

FIG. 15. Probability of triad ABA, P(ABA), versus P(B) for nine sets of reaction probabilities. (See Table 2 for key to sets.)

one would expect, there is more AAB^+ for set 6 from about $P(B) = 0.3$
on. Also, there is less ABB^+ for sets 2, 4, 6, 8 only from about
$P(B) = 0.3$ on, while at lower conversion there is an excess of ABB^+
for 2, 4, 8. One reaches here the limit of the simple explanation
given for the qualitative behavior of sequence probabilities in
terms of block character and alternation. Thus computer experiments
are not only of value to yield quantitative data on conversion and
on sequence probabilities but also to yield difficult to anticipate,
qualitative features of the statistics of triads and probably also
of longer sequences.

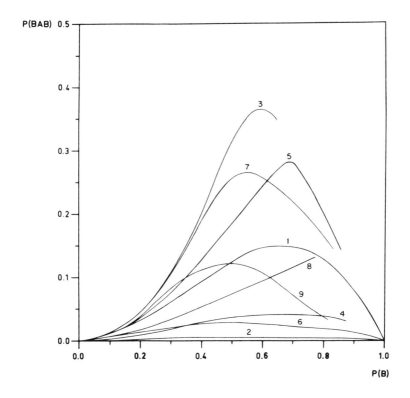

FIG. 16.   Probability of triad BAB, P(BAB), versus P(B) for
nine sets of reaction probabilities. (See Table 2 for key to sets.)

For the simulation of actual reactions on polymers it is of interest in some cases to be able to change the set of reaction probabilities during the computer run. A simulation where the set p(000); p(001^{+}); p(101) = 1.0; 1.0; 1.0 has been changed to 0.01; 0.1; 1.0 is presented by the probabilities of the A-centered triads versus P(B) in Fig. 19. The data of the simulation are shown as points while the drawn-out curves have been calculated for randomness of A and B units, i.e., Bernoullian statistics. The conversion where the set of reaction probabilities has been changed is indicated by arrows. Up to the arrows the data points follow the

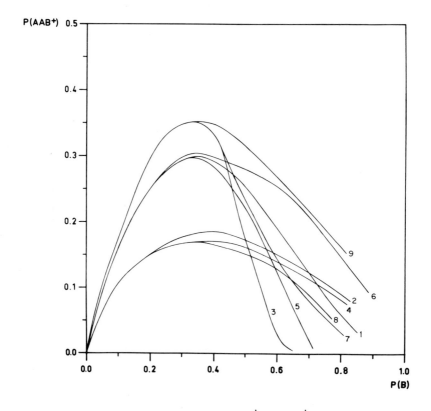

FIG. 17. Probability of triad AAB^{+}, P(AAB^{+}), versus P(B) for nine sets of reaction probabilities. (See Table 2 for key to sets.)

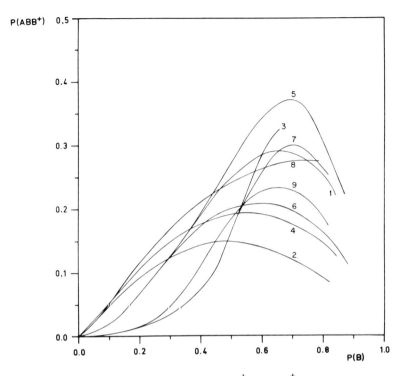

FIG. 18. Probability of triad ABB^+, $P(ABB^+)$, versus $P(B)$ for nine sets of reaction probabilities. (See Table 2 for key to sets.)

Bernoullian curves closely, as they must, but then they deviate immediately to yield a triad distribution of the expected block character.

Monte Carlo simulation is also useful in determining the ratio of rate constants of actual reactions on polymers from measured conversion or sequence data. Toward this end one may change the reaction probabilities by trial and error until a best fit to the measured data is obtained. The simulated curves of Figs. 7, 8, and 10 to 18 may be helpful in this process by giving a first indication of the reaction probabilities which might fit the experimental data. Figure 8 shows, however, that the shape of the $P(A)$ versus log C curves generally does not change as greatly with a change in

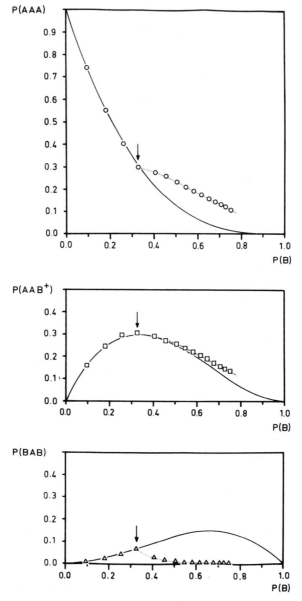

FIG. 19. Probabilities of A-centered triads versus P(B). Triad probabilities which are obtained by simulation are shown as points. The simulation was carried out with set p(000); p(001[+]); p(101) = 1.0; 1.0; 1.0 up to the point indicated by arrows and then with set 0.01; 0.1; 1.0. The calculated Bernoullian triad probabilities are shown as drawn-out curves for comparison.

the ratio of reaction probabilities as the P(XX) versus P(B) and
the P(XXX) versus P(B) curves (X = A,B), if it is realized that the
position of a given P(A) versus log C curve parallel to the log C axis
is independent of the ratio of the reaction probabilities. Thus,
for obtaining better precision, sequence data instead of conversion
data should be fitted. In addition, the shape of curves of P(A)
versus log t of an actual reaction depends on the order of [R].
Since the curves of Fig. 8 are simulated with assumption of zeroth
order in [R], only actual reactions of this order in [R] may be
fitted. For the P(XX) versus P(B) plots, however, the order in [R]
of an actual reaction does not effect the data in these plots[2].

   As the first example, the curve-fitting procedure has been
carried out on measured P(XXX) versus P(B) data obtained during the
hydrolysis of syndiotactic poly(methyl methacrylate) to methyl
methacrylate-methacrylic acid copolymers [42]. The data were obtained by [1]H-nmr spectroscopy. Hydrolysis had to be carried out,
however, in two steps. In the first step the poly(methyl methacrylate) was hydrolyzed to relatively low conversion [P(B) = 0.37] in
an alkaline organic medium which resulted in a copolymer of Bernoullian statistics. This copolymer was, in contrast to the starting poly(methyl methacrylate), soluble in aqueous KOH, in which
medium the second hydrolysis step was performed. With an excess of
KOH in aqueous solution the cooperative effect of neighboring monomer units leads during the second hydrolysis step to a copolymer
with a tendency toward alternation. In keeping with the two step
hydrolysis, the simulation was carried out with a change in the set
of reaction probabilities similar to Fig. 19. The first set was
always the random set which was changed at P(B) = 0.37 to several
sets selected to fit the [1]H-nmr data of the second hydrolysis step.
In Fig. 20, the measured probabilities of the A-centered triads obtained during the second hydrolysis are shown as points; the hydrolysis starts with a random copolymer of P(B) = 0.37 (point fur-
thest to the left). The drawn-out curves in Fig. 20 represent only
the part of the simulation which was performed with the second set
of reaction probabilities. One of the curves was obtained with the

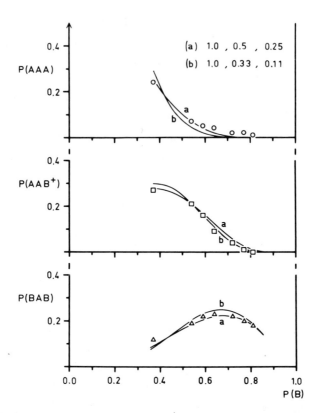

FIG. 20. Probabilities of A-centered triads versus P(B). Triad probabilities obtained by nmr measurement on syndiotactic methyl methacrylate methacrylic acid copolymers are shown as points. Triad probabilities obtained by simulation with sets 1.0; 0.5; 0.25 and 1.0; 0.33; 0.11 (1000 monomer units each) are represented by drawn-out curves.

set 1.0; 0.5; 0.25 and the other with 1.0; 0.33; 0.11. Apparently the former set of reaction probabilities gives a somewhat better fit than the latter. Thus one might state that the ratio of the rate constants for the hydrolysis in aqueous solution with an excess of KOH is approximately 4:2:1.

The above ratio of rate constants responsible for producing alternation may be explained by the electrostatic repulsion between attacking OH⁻ and negatively charged methacrylic-acid units which

are present as next neighbors to the methyl-methacrylate unit
being attacked. This leads to rate constants of the order k(AAA) >
k(AAB$^+$) > k(BAB) in keeping with the increasing electrostatic repulsion. On the other hand, if the hydrolysis is carried out in
aqueous solution with less KOH than is necessary to have all methacrylic-acid units negatively charged, no OH$^-$ is available as the
attacking species. The hydrolysis proceeds nevertheless with the
unneutralized methacrylic-acid units, which is now the attacking
species. Since this species is now bound to the polymer chain, the
next neighboring methyl-methacrylate units are preferably attacked
with the result that copolymers of block character are formed and
that the rate constants are of the order k(AAA) < k(AAB$^+$) < k(BAB)
[42].

III. COMPUTER CALCULATION OF SEQUENCES OF PLACEMENTS IN HOMOPOLYMERS DERIVED BY REACTIONS ON COPOLYMERS

In the present section an example of the use of computer calculations for reactions on polymers based on work of Gronski and
Klesper [6], will be discussed. Computer calculations, as opposed
to Monte Carlo experiments, rely on equations which are processed
with the aid of a computer to obtain the desired result.

When converting a copolymer by a chemical reaction to a homopolymer, the sequencing of the placements, i.e., relative configurations of neighboring monomer units, may be retained in the homopolymer. Thus a vinyl-type binary copolymer prepared by copolymerization and possessing both syndiotactic and isotactic placements
can, in cases, be chemically transformed to a homopolymer built from
only one type of monomer unit. This may be represented in a schematic way on an arbitrary chain segment:

$$\cdots A_i B_s A_i A_i A_i B_s B_s B_i B_s A \cdots$$

$$\downarrow \quad \text{Chemical reaction}$$

$$\cdots C_i C_s C_i C_i C_i C_s C_s C_i C_s C \cdots$$

where s and i are the syndiotactic and isotactic placements between neighboring monomer units, and C is a monomer unit which may, or may not, be different from monomer units A and B, depending on the chemical reaction used. The type of chemical reaction discussed in this section is in a loose sense the reverse of that of the previous section, where the interest focused on the transformation of a homopolymer to a copolymer. However, the kinetics of the reaction is not to be discussed here, but only the sequencing of the placements in the final product which, in fact, is already present in the starting copolymer.

Nuclear magnetic resonance studies of copolymers prepared by copolymerization have shown that the terminal model of copolymerization is frequently sufficient for the complete statistical description of the sequencing of monomer units. According to this model only the last monomer unit of the growing chain has an influence on the kinetics of the copolymerization, not the monomer units further removed from the growing chain end. The probabilities of the sequences of monomer units are then fully characterized by the product of the first order reactivity ratios, $r_A \cdot r_B$. It is often implicitly assumed that the terminal model holds also for the configuration, which is reasonable. Four parameters are then to be defined for placements, σ_{AA}, σ_{AB}, σ_{BA}, and σ_{BB}, whereby, for instance, σ_{AB} is the conditional probability that an isotactic placement is between a given A and B, A and B being neighbors and B having been added to a chain ending in A.

A. Basic Formalism

According to the terminal model of copolymerization the probability of an isotactic dyad is given by

$$P\binom{k\ 1}{i} = P(k)P(k|1)\sigma_{k1} \qquad\qquad k, 1 = A, B \qquad\qquad (8a)$$

and for a syndiotactic dyad it may be written

$$P\binom{k\ 1}{s} = P(k)P(k|1)(1 - \sigma_{k1}) \qquad\qquad\qquad (8b)$$

where k and 1 represent monomer units A or B, and i and s represent isotactic and syndiotactic placements. $P(k)$ is the probability of a monomer unit k, and $P(k|l)$ is the conditional probability for an l unit given a neighboring k unit.

The probability of an isotactic or syndiotactic placement is obtained by summing over all possible isotactic or syndiotactic dyads of monomer units, respectively [57]. For the probability of an isotactic placement:

$$P(i) = \sum_{k,l=A,B} P\binom{k\ l}{i} \tag{9a}$$

For the probability of a syndiotactic placement:

$$P(s) = \sum_{k,l=A,B} P\binom{k\ l}{s} \tag{9b}$$

Equations analogous to Eqs. (9a) and (9b) may also be written for probabilities of sequences of placements, for instance:

$$P(isi) = \sum_{k,l,m,n=A,B} P\binom{k\ l\ m\ n}{i\ s\ i} \tag{10}$$

$$P\binom{k\ l\ m\ n}{i\ s\ i} = P(k)P(k|l)\sigma_{kl}P(l|m)(1 - \sigma_{lm})P(m|n)\sigma_{mn} \tag{11}$$

In connection with Eqs. (8) and (11) it should be noted that the order of writing a sequence, for example, $\begin{smallmatrix}k\ l\ m\ n\\i\ s\ i\end{smallmatrix}$, by one probability and a series of conditional probabilities is immaterial, in keeping with the property of a Markov chain of first order. The combined Eqs. (10) and (11) may also be written as the trace of a matrix product. This notation is more compact and lends itself easily to computer programming:

$$P(i\ s\ i) = \mathrm{Tr}\ \underline{\phi}\ \underline{i}\ \underline{si} \tag{12}$$

The matrices $\underline{\phi}$, $\underline{i}$, and $\underline{s}$ are:

$$\underline{\phi} = \begin{bmatrix} P(A) & P(B) \\ P(A) & P(B) \end{bmatrix} \tag{13a}$$

$$\underline{i} = \begin{bmatrix} P(A|A)\sigma_{AA} & P(A|B)\sigma_{AB} \\ P(B|A)\sigma_{BA} & P(B|B)\sigma_{BB} \end{bmatrix} \tag{13b}$$

$$\underline{s} = \begin{bmatrix} P(A|A)(1 - \sigma_{AA}) & P(A\ B)(1 - \sigma_{AB}) \\ P(B|A)(1 - \sigma_{BA}) & P(B\ B)(1 - \sigma_{BB}) \end{bmatrix} \tag{13c}$$

where $P(A|B)$, for instance, is the conditional probability for a B, given a neighboring A. For any sequence of placements $X_1 X_2 X_3 \cdots X_n$, where X_k is i or s and $k = 1,2,3,\ldots,n$:

$$P(X_1 X_2 X_3 \cdots X_n) = \mathrm{Tr}\ \underline{\phi}\ \underline{X}_1 \underline{X}_2 \underline{X}_3 \cdots \underline{X}_n \tag{14}$$

with $\underline{X}_k = \underline{i}$ or $\underline{s}$, and when $X_k = i$, then $\underline{X}_k = \underline{i}$. The elements $P(k)$ and the factors $P(k|1)$ in the matrices are related via the product of the reactivity ratios $r_A \cdot r_B$:

$$P(A|B) = \frac{1}{1 + r_A F} \tag{15a}$$

$$P(B|A) = \frac{1}{1 + r_B/F} \tag{15b}$$

The feed ratio F is given by the rearranged terminal model copolymerization equation:

$$F = \frac{P(A) - P(B) + \sqrt{[P(A) - P(B)]^2 + 4P(A)P(B)r_A \cdot r_B}}{2P(B)r_A} \tag{15c}$$

In accordance with Eqs. (14) and (15) one may vary $P(A)$ in sufficiently small steps for a given $r_A\ r_B$ and a given set of σ_{kl} to draw a smooth curve through the calculated values of $P(X_1 X_2 X_3 \cdots X_n)$.

B. Statistics of Sequences of Placements

As is obvious from the foregoing, the statistics of the sequences of placements found in a homopolymer is dependent on the statistics of monomer units in the copolymer from which the homopolymer is derived. Three products of reactivity ratios were chosen to cover the main types of sequencing of monomer units in the copolymers, i.e., $r_A \cdot r_B = 1$ for Bernoullian statistics, $r_A \cdot r_B = 0.03$ for a tendency toward alternation, and $r_A \cdot r_B = 10$

for block character. The statistics of the sequences of placements in the copolymers is, of course, also dependent on the σ_{kl}. Six sets of σ_{kl} were chosen which are given in Table 3. It should be noted, however, that most of these sets of σ_{kl} have been chosen to cover extreme cases rarely to be expected in copolymerization. However, the extreme sets have the advantage that the results can be rationalized in a simple way. With respect to Table 3 it should be emphasized that exchanging the numerical values for σ_{AB} and σ_{BA} within a given set will not alter the statistics of the placements [6,57], i.e., $P(X_1X_2X_3\cdots X_n)$. Also, exchanging the numerical values for σ_{AA} and σ_{BB} will not change $P(X_1X_2X_3\cdots X_n)$, provided the numerical values for $P(A)$ and $P(B)$ are interchanged at the same time.

In Fig. 21 plots of $P(s)$ versus $P(B)$ in dependence of $r_A \cdot r_B$ and σ_{kl} are shown. The sets of σ_{kl} are identified by numbers on the curves, the numbers referring to Table 3. Apparently, $P(s)$ depends more on the sets σ_{kl} than on $r_A \cdot r_B$, as one would probably expect. For calculating $P(s)$ a computer is not essential since according to Eqs. (8b) and (9b), $P(s)$ is simply

$$P(s) = P(A)\ P(A|A)\ (1 - \sigma_{AA}) + P(B)\ P(B|A)\ (1 - \sigma_{BA})$$

$$+ P(A)\ P(A|B)\ (1 - \sigma_{AB}) + P(B)\ P(B|B)\ (1 - \sigma_{BB}) \qquad (16)$$

TABLE 3

Sets of σ_{kl} of Copolymers Used for Demonstrating their
Influence on the Statistics of Placements

Set	σ_{AA}	σ_{AB}	σ_{BA}	σ_{BB}
1	0	0	1	1
2	0	0.5	0.5	1
3	1	0	0	1
4	0	1	1	1
5	0.3	0.5	0.5	1
6	0.3	0.5	0.5	0.7

P(s)

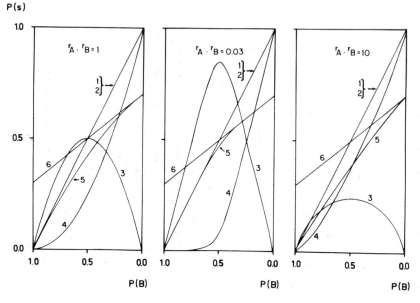

FIG. 21. Probability of syndiotactic placement s, P(s), versus P(A) calculated for six sets of placement parameters σ_{kl} (curves resulting from sets numbered according to Table 3) and three different products of reactivity ratios, $r_A \cdot r_B$.

With the usual normalization equations for probabilities and conditional probabilities and the relation $P(A) \, P(A|B) = P(B) \, P(B|A)$, one may rewrite Eq. (16)

$$P(s) = P(A) \left\{ \sigma_{BB} - \sigma_{AA} + P(A|B) [\sigma_{BB} + \sigma_{AA} - \sigma_{AB} - \sigma_{BA}] \right\}$$
$$+ \ 1 - \sigma_{BB} \qquad (17)$$

For sets 1, 2, and 6 of Table 3 it holds $\sigma_{BB} + \sigma_{AA} - \sigma_{AB} - \sigma_{BA} = 0$. This is reasonable for copolymerization since σ_{AB} and σ_{BA} are likely to possess values which are between σ_{BB} and σ_{AA}. Eq. (17) simplifies for sets 1, 2, and 6 to

$$P(s) = P(A) \ [\sigma_{BB} - \sigma_{AA}] + 1 - \sigma_{BB} \qquad (18)$$

Equation (18) explains the straight line relationship of sets 1, 2, and 6; it explains also the lack of dependence on $r_A \cdot r_B$ since Eq. (18) does not contain the conditional probability $P(A|B)$ of

Eq. (17). Another reasonable condition in copolymerization is $\sigma_{AB} = \sigma_{BA}$, which holds for the remaining sets 3, 4, and 5. Substitution of the values of σ_{kl} for these sets into Eq. (17) leads to

$$P(s) = 2\ P(AB) \qquad\qquad (\text{set } 3) \qquad\qquad (19a)$$

$$P(s) = P(AA) \qquad\qquad (\text{set } 4) \qquad\qquad (19b)$$

$$P(s) = 0.7\ P(A) + 0.3\ P(AB)$$
$$ = 0.7\ P(AA) + P(AB) \qquad (\text{set } 5) \qquad\qquad (19c)$$

The dependence of $P(s)$ on $r_A \cdot r_B$ for the sets 3, 4, and 5 in Fig. (21) is now readily understood because $r_A \cdot r_B = 0.03$ favors formation of AB and disfavors AA in comparison to $r_A \cdot r_B = 1$, while $r_A \cdot r_B = 10$ disfavors AB and favors AA also in relation to $r_A \cdot r_B = 1$.

Turning now to sequences of placements, one might plot directly their probabilities. It is preferred, however, to plot the persistence ratios [58], which give not only information about the tendency for alternation and the block character of the sequencing of placements but also about the lowest order of conditional probabilities useful for calculating longer sequences of placements from shorter ones. In addition they give a measure of the deviation from Bernoullian statistics since the persistence ratios relate the conditional probabilities of placements to a Bernoullian statistics of placements represented by $P(s)$;

$$\zeta_i = \frac{P(s)}{P(i\,|\,s)} \qquad\qquad\qquad\qquad (20a)$$

$$\eta_s = \frac{P(s\,|\,s)}{P(s)} \qquad\qquad\qquad\qquad (20b)$$

$$\eta_{is} = \frac{P(is\,|\,s)}{P(s)} \qquad\qquad\qquad\qquad (20c)$$

$$\eta_{ss} = \frac{P(ss\,|\,s)}{P(s)} \qquad\qquad\qquad\qquad (20d)$$

where $P(is\,|\,s)$, for instance, is the conditional probability of placement s given that the two contiguous placements are i and s in the order iss (or ssi). The persistence ratios of Eqs. (20a) through (20d) provide criteria for determining the Markov order of the statistics of the placements:

$$1 = \zeta_i = \eta_s = \eta_{is} = \eta_{ss} \quad \text{(Bernoullian = Markov}$$
$$\text{zeroth order)} \qquad (21a)$$

$$1 \neq \zeta_i \neq \eta_s = \eta_{is} = \eta_{ss} \neq 1 \quad \text{(Markov first order)} \qquad (21b)$$

$$1 \neq \zeta_i \neq \eta_s \neq \eta_{is} \neq \eta_{ss} \quad \text{(higher order Markovian}$$
$$\text{or non-Markovian)} \qquad (21c)$$

If the persistence ratios ζ or $\eta > 1$, a block character of placements prevails; but when ζ or $\eta < 1$, a tendency toward alternation
is indicated.

The first of the persistence ratios, ζ_i, is plotted versus
$P(s)$ in Fig. 22 as drawn-out curves, while the analog of ζ_i with
respect to monomer units, i.e., $\zeta_B = P(A)/P(B|A)$, is shown as the
dashed curves with the axis of Fig. 22 to be read as ζ_B and $P(A)$,
as indicated. The unequal range of $P(s)$ covered by the drawn-out
curves arises because of the limitations imposed by the sets of
σ_{k1}, the $P(A)$ being varied in all cases from 0.1 to 0.9. The
drawn-out curves coincide with the dashed curves for set 1 of σ_{k1}.
This is due to the exact transcription of the sequencing of monomer
units unto the sequencing of placements for set 1, as is to be discussed later. Because the sequencing of monomer units is Bernoullian for $r_A \cdot r_B = 1$ and Markov first order for $r_A \cdot r_B = 0.03$ and
$r_A \cdot r_B = 10$, it follows that the same must hold true for the sequencing of placements for set 1.

Fig. 22 shows, in view of $\zeta_i > \zeta_B$, that for the other sets of
σ_{k1} and for $r_A \cdot r_B = 1$ and $r_A \cdot r_B = 0.03$ the block character of
the sequencing of placements is more pronounced than the "block
character" of the underlying sequencing of monomer units (if one
wants to speak of "block character" for the random and alternating
statistics of monomer units for these $r_A \cdot r_B$). This tendency to
increased block character of placements is more marked with the
extreme sets 3 and 4 than with the sets 5 and 6 which are more
likely for copolymerization. If the sequencing of monomer units is
not random or alternating but of block character, as with $r_A \cdot r_B =$
10, then the block character of the placements may be either higher

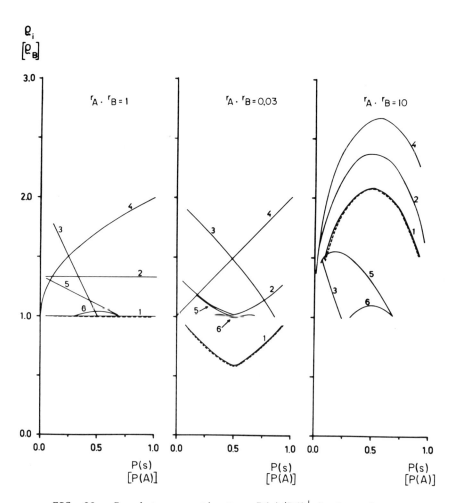

FIG. 22. Persistence ratio $\zeta_i = P(s)/P(i|s)$ plotted versus $P(s)$ (drawn-out curves), and $\zeta_B = P(A)/P(B|A)$ versus $P(A)$ (dashed curves). Calculation of ζ_i, carried out for six sets of σ_{kl} (drawn-out curves resulting from sets numbered according to Table 3).

or lower than the block character of the monomer units, depending on the set of σ_{kl}. At any $r_A \cdot r_B$, it is, however, difficult to produce a tendency for alternation of the placements which is obvious from the fact that usually $\zeta_i > 1$.

For all sets σ_{kl} and all $r_A \cdot r_B$, in general, $\zeta_i \neq 1$, which means that according to Eq. (21b) at least first order Markovian statistics prevails. The only exception is set 1 at $r_A \cdot r_B = 1$, as already mentioned. Plotting now in Fig. 23 the persistence ratios η in addition to ζ_i for one selected set (set 2) shows that in fact even a first-order Markov statistics does not hold. This is clear from the difference between η_s, η_{is}, and η_{ss} (curves 2, 3, and 4 in Fig. 23) and Eq. (21b). This result may be somewhat surprising because the terminal model in copolymerization must always lead to Markov first order for the sequencing of monomer units. Moreover, the differences between the η are usually not small. Thus even an approximate calculation of longer sequences of placements from first-order conditional probabilities is not feasible in general with set 2.

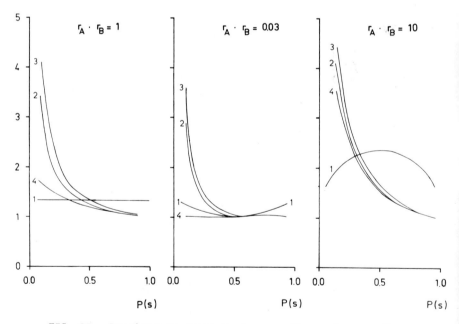

FIG. 23. Persistence ratio ζ_i (curves 1), η_s (curves 2), η_{is} (curves 3), and η_{ss} (curves 4), as defined by Eqs. (20), are plotted versus $P(s)$ for one set of σ_{kl} (set 2 of Table 3) and three $r_A \cdot r_B$.

Now it can be shown that the statistics of placements in homopolymers derived from a terminal model copolymer should be non-
Markovian in general. Writing the condition for non-Markovian behavior as

$$P\left(\frac{X_m X_{m+1} \cdots X_{m+n-1} X_{m+n}}{X_m X_{m+1} \cdots X_{m+n-1}}\right) \neq P\left(\frac{X_{m-1} X_m X_{m+1} \cdots X_{m+n-1} X_{m+n}}{X_{m-1} X_m X_{m+1} \cdots X_{m+n-1}}\right)$$

$$(X = i, s; m, n = 1, 2, 3, \ldots) \qquad (22)$$

and writing the traces of the matrix products for Eq. (22) in accordance with Eq. (14) one obtains:

$$\frac{\text{Tr } \phi \underline{X}_m \underline{X}_{m+1} \cdots \underline{X}_{m+n-1} \underline{X}_{m+n}}{\text{Tr } \phi \underline{X}_m \underline{X}_{m+1} \cdots \underline{X}_{m+n-1}} \neq \frac{\text{Tr } \phi \underline{X}_{m-1} \underline{X}_m \underline{X}_{m+1} \cdots \cdots \underline{X}_{m+n-1} \underline{X}_{m+n}}{\text{Tr } \phi \underline{X}_{m-1} \underline{X}_m \underline{X}_{m+1} \cdots \cdots \underline{X}_{m+n-1}}$$

$$(\underline{X} = \underline{i}, \underline{s}) \qquad (23)$$

It is evident from the matrix notation that in general the inequal-
ity sign of Eqs. (22) and (23) is justified. The non-Markovian behavior arises because of the sharing of one monomer unit by adjacent
placements which creates a stochastic dependence of contiguous
placements as soon as the σ_{kl} of adjacent placements differ. The
non-Markovian behavior disappears correspondingly when the σ_{kl} are
the same, i.e., if $\sigma_{AA} = \sigma_{AB} = \sigma_{BA} = \sigma_{BB}$.

The influence of $r_A \cdot r_B$ and σ_{kl} on the sequencing of placements may also be studied by considering run numbers. Harwood [59]
defined a run number for the sequencing of monomer units

$$R = 100[P(AB) + P(BA)] \qquad (24)$$

which is the average number of the sum of blocks of A units and
blocks of B units per 100 monomer units. A run number for the sequencing of placements may be defined analogously:

$$R_{conf} = 100[P(is) + P(si)] \qquad (25)$$

where R_{conf} is the average number of the sum of blocks of syndiotactic placements and blocks of isotactic placements per 100 placements. In Fig. 24 the drawn-out curves give R_{conf} plotted versus
P(s) and the dashed curves R versus P(A). A small run number

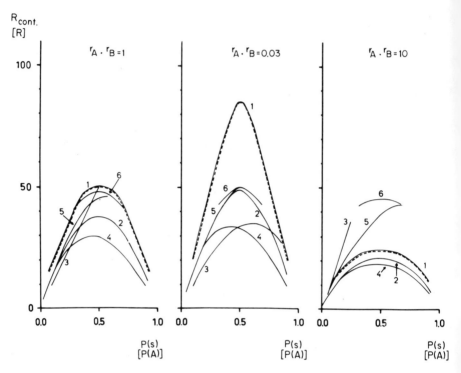

FIG. 24. Run number of placements, R_{conf}, versus P(s) shown
as drawn-out curves. Run number of monomer units, R, versus P(A)
shown as dashed curves. R_{conf} calculated for six sets of σ_{kl}.

indicates block character whereby as point of reference one may conveniently choose the dashed curve of $r_A \cdot r_B = 1$ which, of course,
represents a Bernoullian statistics for R. Moreover, this dashed
curve represents a Bernoullian statistics for R_{conf} because the
drawn-out curve of the set of exact transcription, set 1, superim-
poses on the dashed curve. Comparing the drawn-out curves of all
$r_A \cdot r_B$ with this reference curve, it is obvious that the sequencing
of placements usually has block character which corroborates Fig. 22.

Comparing now the drawn-out and dashed curves within each plot
of a given $r_A \cdot r_B$, the conclusion of Fig. 22 is verified, that the
block character of the sequencing of placements is more pronounced
than the "block character" of the underlying sequencing of monomer

units as long as the underlying sequencing is random or alternating $(r_A \cdot r_B = 1$ and $r_A \cdot r_B = 0.03)$. When the underlying sequencing is of block character $(r_A \cdot r_B = 10)$, there exist sets of σ_{kl} which enhance the block character and others which reduce it. The latter fact is also reflected in the average block lengths of placements. The average block length of s placements is given by [60]:

$$\overline{n}_s = \frac{P(s)}{P(si)} \tag{26}$$

and that of i placements by:

$$\overline{n}_i = \frac{P(i)}{P(is)} = \frac{1 - P(s)}{P(si)} \tag{27}$$

In Fig. 25 $\overline{n}_i$ is plotted versus $P(s)$ for $r_A \cdot r_B = 10$. Comparison with set 1 shows that the average block length of i placements is increased for sets 2 and 4 and decreased for sets 3, 5, and 6. Thus the behavior of these sets in Fig. 24 carries over also to specifically blocks of i placements.

For the qualitative rationalization of the behavior of R_{conf} in Fig. 24, one conveniently considers the individual contributions of the blocks of monomer units A and of blocks of monomer units B to the number of blocks of placements. Considering set 1 first, every block of monomer units AB^nA and BA^nB (n = 1, 2, 3, ...) has in view of $\sigma_{AA} = 0$, $\sigma_{AB} = 0$, and $\sigma_{BB} = 1$, $\sigma_{BA} =$ one block of si^ns, and is^ni, respectively, associated with it. Because every copolymer can be visualized as consisting only of blocks AB^nA and BA^nB, there is therefore an exact transcription of the sequencing of monomer units onto the sequencing of placements, and R and R_{conf} coincide with set 1 regardless of $r_A \cdot r_B$. For set 2 there exists in view of $\sigma_{AA} = 0$ and $\sigma_{BB} = 1$ for every AB^nA and BA^nB still one block of si^ms and is^mi as long as n = 2, 3, 4, ... (n = m - 1, m, m + 1). If, however, n = 1 there is in view of $\sigma_{AB} = \sigma_{BA} = 0.5$ no longer one block of placements present for each block of monomer units. Thus the number of blocks of placements may be either the same or less than the number of blocks of monomer units $(R_{conf} \leqslant R)$. The difference between R_{conf} and R increases with the increase in the number of

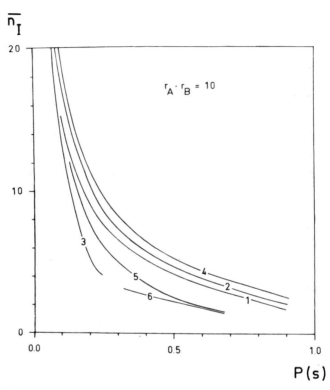

FIG. 25. Average block length of isotactic placements, $\overline{n_i}$, versus P(s) for six sets of σ_{kl} and one $r_A \cdot r_B$ ($r_A \cdot r_B = 10$).

blocks of monomer units for which $n = 1$. Therefore, we observe in Fig. 24 a much smaller R_{conf} with $r_A \cdot r_B = 0.03$, a smaller R_{conf} with $r_A \cdot r_B = 1$, and a little smaller R_{conf} with $r_A \cdot r_B = 10$, all in comparison with the R of the applicable $r_A \cdot r_B$. For set 3 one obtains in view of $\sigma_{AA} = \sigma_{BB} = 1$ for every AB^nA and BA^nB a block of si^ms as long as $n = 2, 3, 4, \ldots (n = m + 1)$. If $n = 1$, a sequence ss is created for each AB^nA and BA^nB because of $\sigma_{AB} = \sigma_{BA} = 0$. This amounts to $R_{conf} = 2R$ when $n = 2, 3, 4, \ldots$ for all blocks of monomer units and $R_{conf} = 1$ when $n = 1$ for all blocks of monomer units. Thus $R_{conf} > R$ with $r_A \cdot r_B = 10$, $R_{conf} < R$ with $r_A \cdot r_B = 0.03$, and $R_{conf} \approx R$ with $r_A \cdot r_B = 1$. For the set 5 which, besides set 6, is more realistic for copolymerization, the special feature arises that for each BA^nB there may occur more than one block of placements as long as $n \geq 3$ because of $0 < \sigma_{AA} > 1$. Similarly, for set 6 there may occur for both AB^nA and BA^nB more than one block of placements each if $n \geq 3$. This then leads for sets 5 and 6 to $R_{conf} > R$ for $r_A \cdot r_B = 10$. For $n < 3$ only one block of placements or less may occur with sets 5 and 6 for each block of monomer units, thus $R_{conf} < R$ for $r_A \cdot r_B = 0.03$.

REFERENCES

1. A. D. Litmanovich, N. A. Platé, O. V. Noah, and V. I. Golyakov, *European Pol. J.*, Supplement, 1969, 517.

2. E. Klesper, W. Gronski, and V. Barth, *Makromol. Chem.*, *150*, 223 (1971).

3. B. N. Goldstein, A. N. Goryunov, Yu. Ya. Gotlieb, A. M. Elyashevich, T. P. Zubova, A. I. Koltzov, V. D. Nemirovskii, and S. S. Skorokhodov, *J. Pol. Sci.*, A-2, *9*, 769 (1971).

4. M. E. Craig and D. M. Crothers, *Biopolymers*, *6*, 385 (1968).

5. N. A. Platé and A. D. Litmanovich, XXIIIrd International Congress of Pure and Applied Chemistry, Butterworths, London, *8*, 123 (1971).

6. W. Gronski and E. Klesper, *J. Pol. Sci.*, *Pol. Phys. Ed.*, *11*, 1963 (1973).

7. A. Silberberg and R. Simha, *Biopolymers*, *6*, 479 (1968).

8. P. Rabinowitz, A. Silberberg, R. Simha, and E. Loftus, in Stochastic Processes in Chemical Physics, vol. 15 of *Adv. Chem. Phys.*, Interscience, New York, 1969, p. 281.

9. H. K. Frensdorff and O. Ekiner, *J. Pol. Sci.*, A-2, *5*, 1157 (1967).

10. N. W. Johnston and H. J. Harwood, *J. Pol. Sci.*, C, *22*, 591 (1969).

11. N. Goodman and H. Morawetz, *J. Pol. Sci.*, C, *31*, 177 (1970).

12. N. Goodman and H. Morawetz, *J. Pol. Sci.* A-2, *9*, 1657 (1971).

13. M. Sisido, *Pol. J.*, *3*, 84 (1972).

14. H. J. Harwood, Y. Kodaira, and D. L. Newman, this volume.

15. H. J. Harwood, *ACS Polymer Preprints*, *12/2*, 46 (1971).

16. N. A. Platé, IUPAC International Symposium on Macromolecules, Budapest 1969, Publishing House of the Hungarian Academy of Sciences, p. 651.

17. J. B. Keller, *J. Chem. Phys.*, *37*, 2584 (1962).

18. J. B. Keller, *J. Chem. Phys.*, *38*, 325 (1963).

19. C. B. Arends, *J. Chem. Phys.*, *38*, 322 (1963).

20. T. Alfrey and W. G. Lloyd, *J. Chem. Phys.*, *38*, 318 (1963).

21. L. Lazare, *J. Chem. Phys.*, *39* 727 (1963).

22. E. R. Cohen and H. Reiss, *J. Chem. Phys.*, *38*, 680 (1963).

23. D. A. McQuarrie, J. P. McTague, and H. Reiss, *Biopolymers, 3*, 657 (1965).

24. N. Gö, *J. Phys. Soc. Japan, 22*, 413 (1967).

25. F. W. Schneider and P. K. Rawlings, *J. Chem. Phys.*, *55*, 1257 (1971).

26. G. Schwarz, *Ber. Bunsenges.*, *76*, 373 (1972).

27. R. Simha and R. H. Lacombe, *J. Chem. Phys.*, *55*, 2936 (1971).

28. R. H. Lacombe and R. Simha, *J. Chem. Phys.*, *58*, 1043 (1973).

29. A. Silberberg and R. Simha, *Macromolecules, 5*, 332 (1972).

30. H. Morawetz and P. E. Zimmering, *J. Phys. Chem.*, *58*, 753 (1954).

31. H. Morawetz and E. W. Westhead, *J. Pol. Sci.*, *16*, 273 (1955).

32. G. Smets and A. M. Hesbain, *J. Pol. Sci.*, *40* 217 (1959).

33. W. DeLoecker and G. Smets, *J. Pol. Sci.*, *40*, 203 (1959).

34. G. Smets and W. DeLoecker, *J. Pol. Sci.*, *41*, 375 (1959).

35. G. Smets and W. DeLoecker, *J. Pol. Sci.*, *45*, 461 (1960).

36. J. Semen and J. B. Lando, *Macromolecules, 2*, 570 (1969).

37. F. C. Baines and J. C. Bevington, *J. Pol. Sci.*, A-1, *6*, 2433 (1968).

38. J. C. Bevington and J. R. Ebdon, *Makromol. Chem.*, *153*, 165 (1972).

39. M. Higuchi and R. Senju, *Pol. J.*, *3*, 570 (1972).

40. K. Fujii, S. Brownstein, and A. M. Eastham, *J. Pol. Sci.*, A-1, *6*, 2377 (1968).

41. A. B. Robertson and H. J. Harwood, *ACS Polymer Preprints, 12*, 620 (1971).

42. E. Klesper, W. Gronski, and V. Barth, *Makromol. Chem.*, *139*, 1 (1970).

43. C. David, A. De Pauw, and G. Geuskens, *J. Pol. Sci.*, C, *22,*319 (1968).

44. E. Klesper, D. Strasilla, and V. Barth in "Reactions on Polymers" (J. A. Moore, ed.), D. Reidel Publishing Company, Dortrecht, Holland, 1973, p. 137.

45. N. A. Plate, *ibid.*, p. 169

46. H. J. Harwood, *ibid.*, p. 188.

47. M. Kolínský, D. Doskočilová, B. Schneider, and J. Štokr, *J. Pol. Sci.*, A-1, *9*, 791 (1971).

48. V. R. Allen and R. D. Young, *ACS Polymer Preprints, 10,* 753 (1969).

49. F. Klesper, D. Strasilla, and W. Regel, *Makromol. Chem., 175/2,* 523 (1974).

50. D. Strasilla and E. Klesper, *Makromol. Chem., 175/2,* 535 (1974).

51. E. Klesper, *J. Pol. Sci.*, B, *6*, 315 (1968).

52. E. Klesper, *J. Pol Sci.*, B, *6*, 663 (1968).

53. W. Sliwka, *Angew. Makromol. Chem., 4/5.* 310 (1968).

54. E. Klesper, V. Barth, and A. Johnsen, *XXIIIrd International Congress of Pure and Applied Chemistry*, Butterworths, London, *8*, 151 (1971).

55. H. J. Harwood, private communication.

56. E. Klesper, A. Johnsen, and W. Gronski, *Makromol. Chem., 160,* 167 (1972).

57. E. Klesper, *J. Pol. Sci.*, A-1, *8* 1191 (1970).

58. M. Reinmöller and T. G. Fox, *ACS Polymer Preprints, 7,* 987 (1966).

59. H. J. Harwood and W. M. Ritchey, *J. Pol. Sci.*, B, *2*, 601 (1964).

60. E. Klesper and G. Sielaff in "Polymer Spectroscopy", (D. O. Hummel, ed.), *Verlag Chemie*, Weinheim, 1974, p. 189.

Chapter 2

STOCHASTIC CALCULATIONS OF POLYMER STRUCTURE

H. James Harwood

Yasuto Kodaira

Daniel L. Newman

Institute of Polymer Science
The University of Akron
Akron, Ohio

I. INTRODUCTION

Studies on the physical and chemical properties of tactic polymers, copolymers, terpolymers, etc., and studies on polymerization
mechanisms often involve calculations of the relative amounts of
structural features present in polymers. The computer is a great
asset for performing such calculations for the following reasons:

1. A large number of different quantities must be calculated in
   some cases. In the case of a copolymer derived from symmetrical monomers (e.g., A and B), there are three experimentally
   distinguishable dyads (AA, AB + BA, and BB), six distinguishable triads (AAA, BAA + AAB, BAB, BBB, ABB + BBA, and ABA),
   ten tetrads, and twenty pentads. Even larger numbers of
   structural aspects must be considered in the case of terpolymers derived from symmetrical monomers, or in the case of copolymers derived from monomers that can adopt several configurations when incorporated into polymer chains.
2. The calculations involved are often complex. The equations
   required for a given problem are easy to set up, but their
   solution requires considerable algebraic manipulation, and
   the solutions obtained often have a complicated form. Typo-
   graphical errors are common in published solutions to such
   problems and their form is conducive to calculation error.
   The form of the solutions obtained is such that they provide
   little or no insight about polymerization; they are useful
   only because of the numerical answers they can provide. The
   computer can be programmed to calculate structural features
   from the equations developed by the algebraic process, but
   this is not the only way it can be helpful; it can also be
   programmed to calculate structural features of polymers
   without resorting to complicated, tedious, algebraic manipulation.
3. Most theoretical treatments of copolymerization, terpolymerization, etc., are concerned only with the instantaneous
   behavior of polymerization systems. They are applicable to
   polymers prepared in low conversion, or under conditions in
   which the composition of the monomer mixture does not drift
   with conversion, but they are not generally applicable to
   polymers perpared in high conversion. With the exception of
   the terminal model for copolymerization, it has not been
   possible to derive integrated forms of copolymer equations.
   Numerical integration, done with the aid of a computer,
   provides a convenient solution to this problem.

Elsewhere in this volume, Monte Carlo methods for performing
polymer structure calculations are described. The Monte Carlo
method is easy to apply and is very versatile, but it is time

consuming and costly since many polymer chain realizations must occur before a statistically reliable result is obtained. For this reason, the direct calculation [1-78] of structural features is preferred when possible.

This chapter describes a general computer-oriented procedure for calculating structural features of polymers derived from a variety of polymerization mechanisms. The method is based on theoretical contributions by Coleman [7,25,25] and Price [24,67] and on numerous extensions of their treatments that have been provided by others. An effort has been made to derive from these treatments a set of easy to follow instructions for programming the calculation of structural aspects of polymers. The computer oriented procedure described requires very little mathematical manipulation and can be applied to a large variety of problems. It is illustrated by several applications and a sample program.

II. DEFINITIONS AND CONCEPTS

The structural features of interest in a copolymer derived from monomers A and B include the relative amounts of A and B units, the relative amounts of A-A, B-A, A-B, and B-B pairs (termed dyads), the relative amounts of AAA, BAB, BBB, etc., sequences (triads), and so forth. We will refer to such quantities as monomer-unit concentrations, and as dyad, triad, tetrad, or pentad distributions, and will represent such quantities by the symbol P( ), where the parentheses enclose the structural feature of interest. Such quantities will be considered to be normalized such that

$P(A) + P(B) = 1$

$P(AA) + P(AB) + P(BA) + P(BB) = 1$

$P(AAA) + P(AAB) + P(BAA) + P(BAB) + P(BBB) + P(BBA)$
$$+ P(ABB) + P(ABA) = 1$$

$\Sigma P(XXXX) = 1 \quad (X = A \text{ or } B)$

$\Sigma P(XXXXX) = 1 \quad (X = A \text{ or } B)$

The various distributions are thus *unconditional probabilities* of finding the various structures at any point in the copolymer chain.

Such probabilities are referred to as *stationary concentrations* by many authors.

If a copolymer chain is considered to be infinitely long, or cyclic, so that effects of chain ends can be neglected, there are a number of simple stoichiometric relationships among the unconditional probabilities. These are discussed at length by Coleman and Fox [25,26] and by Ito and Yamashita [35]. A few such relationships are written below.

$$P(A) = P(AA) + P(AB) = P(AA) + P(BA)$$
$$P(B) = P(BB) + P(AB) = P(BB) + P(AB)$$

so that

$$P(AB) = P(BA)$$
$$P(AA) = P(AAA) + P(BAA) = P(AAA) + P(AAB)$$
$$P(BA) = P(ABA) + P(BBA) = P(BAA) + P(BAB)$$
$$P(AAA) = P(AAAA) + P(AAAB)$$
$$P(AABA) = P(AABAA) + P(AABAB)$$

It is thus clear that if pentad distributions are known, tetrad, triad, dyad distributions and monomer concentrations can be calculated from them by addition. The probability of any sequence can be obtained by adding the probabilities of appropriate larger sequences.

The symbol $P(X/Y)$ will be used to represent the *conditional probability* that a unit X follows a given structure Y in the chain. The conditional probabilities are calculated from kinetic parameters and are used to calculate unconditional probabilities. They are normalized such that

$$P(A/A) + P(B/A) = 1$$
$$P(B/B) + P(A/B) = 1$$
$$P(A/AA) + P(B/AA) = 1$$

etc.

It is very easy to calculate structural features of polymers from conditional probabilities and unconditional probabilities, as the following examples illustrate.

$$P(AA) = P(A)P(A/A)$$

$$P(AAB) = P(A)P(A/A)P(B/A) = P(AA)P(A/A)$$

$$P(BAAB) = P(B)P(A/B)P(A/A)P(B/A)$$

It is more difficult, however, to express unconditional probabilities solely in terms of conditional probabilities, and this is
the complicating feature of polymer structure calculations. Once
unconditional probabilities have been evaluated for a given structural aspect (monomer concentrations, dyad distributions, etc.),
the calculation of other structural aspects is done in a straightforward way.

The structural features of interest in a homopolymer containing
pseudoasymmetric atoms, such as poly(methyl methacrylate), include
the mole fractions of *meso*(m) and *racemic*(r) monomer-pair placements,
the relative amounts of mm, mr, rm, and rr stereosequences, etc.

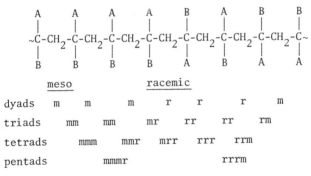

	meso			racemic			
dyads	m	m	m	r	r	r	m
triads		mm	mm	mr	rr	rr	rm
tetrads		mmm	mmr	mrr	rrr	rrm	
pentads			mmmr			rrrm		

As in the case of a copolymer, unconditional and conditional probabilities are used in calculating such structural aspects. Conditional probabilities, such as $P(r/m)$, $P(m/r)$, $P(m/mm)$, etc., are
used to evaluate unconditional probabilities such as $P(m)$, $P(mmr)$,
etc.

When stereochemical aspects of copolymers are of interest, the
various structural features are calculated in terms of conditional
probabilities that reflect both structural and configurational
placements. Thus, the unconditional probability of the structure

```
     H     H     A     B
     |     |     |     |
   -C-CH2-C-CH2-C-CH2-C-
     |     |     |     |
     A     A     H     H
        m     r     m
```

could be calculated as $P(A)P_m(A/A)P_r(A/A)P_m(B/B)$, where the sub-
scripts indicate that the conditional placements are of the meso
(m) or racemic (r) type. Most workers define separate conditional
probabilities for structural or configurational placements; viz.,

$$P_m(A/A) = P(A/A) \cdot \sigma_{AA}$$

where σ_{AA} is the conditional probability that an A-A placement is
the conditional probability that an A-A placement is of the meso
type.

The number of conditional probabilities that must be manipulated to calculate features of polymer structure depends on the
complexity of the polymerization model under consideration. In the
case of copolymerizations that do not require considerations of
configurational placements, a single parameter is needed to define
polymer structure when the nature of the growing chain end does not
influence the propagation process (Bernoullian or zero-order Markovian statistics). When the nature of the terminal unit on the
propagating chain is important (terminal model for copolymerization), the conditional probabilities take the form $P(A/A)$, $P(B/B)$,
etc., and two independent ones are required for the calculations.
The structures of copolymers derived from such processes are said
to obey first-order Markovian statistics. If the nature of the two
units at the end of a growing polymer chain influence the process
(penultimate model), the conditional probabilities take the form
$P(A/AA)$, $P(A/BA)$, etc., and four independent quantities are required for the calculations. The structures of copolymers derived
from such processes are said to obey second-order Markovian statistics, and so forth.

As an example of how conditional probabilities are calculated,
consider a copolymerization process in which a penpenultimate

process operates. Such a process would involve sixteen separate
propagation steps, two of which would be the following:

$$\sim\sim\sim AAA\cdot \; + \; A \; \xrightarrow{\;k_1\;} \; \sim\sim\sim AAA\cdot \qquad\qquad (1)$$

$$\sim\sim\sim AAA\cdot \; + \; B \; \xrightarrow{\;k_2\;} \; \sim\sim\sim AAB\cdot \qquad\qquad (2)$$

If ($\sim\sim\sim AAB\cdot$) and (A) and (B) are radical and monomer concentrations,
respectively, and if k_1 and k_2 are the specific rate constants of
these reactions, P(A/AAA) can be calculated by comparing the rate of
reaction (1) to the sum of the rates of reactions (1) and (2)

$$P(A/AAA) = \frac{k_1 \, (\sim\sim\sim AAA\cdot)\,(A)}{k_1(\sim\sim\sim AAA\cdot)\,(A) \; + \; k_2(\sim\sim\sim AAA\cdot)\,(B)}$$

$$= \frac{1}{1 \; + \; k_2(B)/k_1(A)}$$

Conditional probabilities are easily evaluated from rate constants
and monomer concentrations for most polymerization models, regardless of their complexity. Whatever difficulty is encountered in
calculating structural features of polymers derived from complex
processes stems from the difficulty of evaluating unconditional
probabilities from the large number of conditional probabilities
that are involved.

III. THE GENERAL METHOD

In general, four steps are followed in calculating the composition and other structural aspects of a polymer. A fifth step is
sometimes employed when it is convenient to use an artificial model
for the polymerization process. These steps will be outlined in
the present section and will then be illustrated by specific applications in the following section.

A. Step One

The reactions involved in the polymerization model under study
are written, along with appropriate rate expressions. Expressions

for the various conditional probabilities are then derived by comparing rate expressions. The conditions upon which the conditional
probabilities are based may be taken as the different types of
propagating species involved.

Terminal model

$$\sim\sim\sim A\bullet \ + \ A \xrightarrow{\ k_1\ } \ \sim\sim\sim A\bullet$$

$$\sim\sim\sim A\bullet \ + \ B \xrightarrow{\ k_2\ } \ \sim\sim\sim B\bullet$$

$$P(B/A) \ = \ \frac{k_2(\sim\sim\sim A\bullet)(B)}{k_1(\sim\sim\sim A\bullet)(A) \ + \ k_2(\sim\sim\sim A\bullet)(B)}$$

$$= \ \frac{1}{1 \ + \ k_1(A)/k_2(B)}$$

Penultimate model

$$\sim\sim\sim AB\bullet \ + \ A \xrightarrow{\ k_3\ } \ \sim\sim\sim BA\bullet$$

$$\sim\sim\sim AB\bullet \ + \ B \xrightarrow{\ k_4\ } \ \sim\sim\sim BB\bullet$$

$$P(B/AB) \ = \ \frac{1}{1 \ + \ k_3(A)/k_4(B)}$$

B. Step Two

Unconditional probabilities are calculated from the conditional probabilities derived in Step one. The type of unconditional
probability calculated depends on the type of conditional probability evaluated in Step one. Monomer concentrations are evaluated in
the case of the terminal model and dyad distributions are evaluated
in the case of the penultimate model, for example. Without exception, this step is used to evaluate unconditional probabilities for
the conditions (monomer concentrations, dyad distributions, etc.)
imposed in setting up the conditional probabilities.

This step is the most difficult of the five. We will illustrate three methods for accomplishing it in the next section. Two
of these involve algebraic manipulation and are presented as a

guide for understanding previously published methods. The third method is almost completely numerical and is ideally suited for computer usage.

C. Step Three

Unconditional probabilities of entities smaller than those evaluated in Step two are obtained by addition, using stoichiometric (stationary) relationships such as are described in Sec. II. Thus, monomer probabilities are obtained by adding appropriate unconditional dyad probabilities, whereas dyad probabilities are obtained by adding appropriate triad probabilities, and so forth. The required stoichiometric relationships can be written directly, without derivation, by considering the various ways the larger entities can be derived from the smaller ones. Thus, in the case of a terpolymer of A, B, and C, the expressions written below represent simple statements that an AA dyad must be followed (or preceded) by either an A, B, or C unit. The sum of unconditional probabilities of the resulting triads must thus equal the unconditional probability of the dyad.

$$P(AA) = P(AAA) + P(AAB) + P(AAC)$$
$$= P(AAA) + P(BAA) + P(CAA)$$

D. Step Four

Unconditional probabilities of entities larger than those evaluated in Step 2 are obtained by multiplying conditional and unconditional probabilities. This step is also very easy and the required relationships can usually be written by inspection; viz.,

$$P(AA) = P(A)P(A/A)$$
$$P(AAA) = P(A)P(A/A)P(A/A) \text{ or } P(AA)P(A/AA)$$
$$P(AAB) = P(A)P(A/A)P(B/A) \text{ or } P(AA)P(B/AA)$$
$$P(BAAB) = P(B)P(A/B)P(A/A)P(B/A) \text{ or } P(BA)P(A/BA)P(B/AA)$$

In this step, number distribution (N.D.) and weight distribution (W.D.) of sequences are also evaluated, e.g.,

$$\text{N.D.}(A_n) = P(A/A)^{n-1}[1-P(A/A)]$$

or

$$= P(B/BA) \text{ for } n = 1$$

$$= P(A/BA)P(A/AA)^{n-2}P(B/AA) \text{ for } n > 1$$

$$\text{W.D.}(A_n) = nP(A/A)^{n-1}[1-P(A/A)]^2$$

E. Step Five

In special cases, the calculation is set up so that the unconditional probabilities evaluated in the first four steps are artificial. When this is the case, realistic probabilities are calculated in Step five from the artificial ones. Thus, in considering
copolymers in which monomer complexes participate as reactive
species, it is convenient to treat such processes as multicomponent
polymerizations involving monomers A and B and complexes C = (AB)
and D = (BA). Once the unconditional probabilities of entities
containing these species are calculated in steps one through four,
the results are adjusted so that the C and D units are converted to
AB and BA pairs and structural aspects of the copolymer containing
only A and B units are determined. This step is illustrated in the
next section.

IV. APPLICATIONS

A. Copolymerization

We have previously written a program based on Steps one through
four to calculate monomer concentrations, dyad, triad, tetrad, and
pentad distributions, as well as number and weight distributions of
A and B sequences, for copolymers prepared by terminal or penultimate model copolymerization mechanisms. The reader is referred to
an earlier paper for details concerning this application [50].

B. Terminal Model Terpolymerization

1. Step One

The following propagation steps are assumed in the terminal (first-order Markovian) model for terpolymerization. A, B, and C represent monomer units; $\sim\sim\sim A\cdot$, $\sim\sim\sim B\cdot$, and $\sim\sim\sim C\cdot$ represent propagating chain ends; and specific rate constants for the individual reactions are designated k_{AA}, k_{AB}, etc.

$$\sim\sim\sim A\cdot + A \xrightarrow{\ k_{AA}\ } \sim\sim\sim A\cdot$$

$$\sim\sim\sim A\cdot + B \xrightarrow{\ k_{AB}\ } \sim\sim\sim B\cdot$$

$$\sim\sim\sim A\cdot + C \xrightarrow{\ k_{AC}\ } \sim\sim\sim C\cdot$$

$$\sim\sim\sim B\cdot + A \xrightarrow{\ k_{BA}\ } \sim\sim\sim A\cdot$$

$$\sim\sim\sim B\cdot + B \xrightarrow{\ k_{BB}\ } \sim\sim\sim B\cdot$$

$$\sim\sim\sim B\cdot + C \xrightarrow{\ k_{BC}\ } \sim\sim\sim C\cdot$$

$$\sim\sim\sim C\cdot + A \xrightarrow{\ k_{CA}\ } \sim\sim\sim A\cdot$$

$$\sim\sim\sim C\cdot + B \xrightarrow{\ k_{CB}\ } \sim\sim\sim B\cdot$$

$$\sim\sim\sim C\cdot + C \xrightarrow{\ k_{CC}\ } \sim\sim\sim C\cdot$$

In accord with the convention employed in considerations of copolymerization, the following reactivity ratios are defined:

$$r_{AB} = \frac{k_{AA}}{k_{AB}} \quad r_{BA} = \frac{k_{BB}}{k_{BA}} \quad r_{BC} = \frac{k_{BB}}{k_{BC}}$$

$$r_{AC} = \frac{k_{AA}}{k_{AC}} \quad r_{CA} = \frac{k_{CC}}{k_{CA}} \quad r_{CB} = \frac{k_{CC}}{k_{CB}}$$

If (A), (B), and (C) represent the mole fractions of monomers

A, B, and C in the polymerization mixture, the following conditional probability expressions can be derived:

$$P(A/A) = \frac{k_{AA}(\sim\sim A\cdot)(A)}{k_{AA}(\sim\sim A\cdot)(A) + k_{AB}(\sim\sim A\cdot)(B) + k_{AC}(\sim\sim A\cdot)(C)}$$

$$= \frac{1}{1 + (B)/r_{AB}(A) + (C)/r_{AC}(A)}$$

$$P(B/A) = \frac{k_{AB}(\sim\sim A\cdot)(B)}{k_{AA}(\sim\sim A\cdot)(A) + k_{AB}(\sim\sim A\cdot)(B) + k_{AC}(\sim\sim A\cdot)(C)}$$

$$= \frac{(B)/r_{AB}(A)}{1 + (B)/r_{AB}(A) + (C)/r_{AC}(A)}$$

$$P(C/A) = 1 - P(A/A) - P(B/A)$$

$$P(B/B) = \frac{1}{(A)/r_{BA}(B) + 1 + (C)/r_{BC}(B)}$$

$$P(A/B) = \frac{(A)/r_{BA}(B)}{(A)/r_{BA}(B) + 1 + (C)/r_{BC}(B)}$$

$$P(C/B) = 1 - P(B/B) - P(A/B)$$

$$P(C/C) = \frac{1}{(A)/r_{CA}(C) + (B)/r_{CB}(C) + 1}$$

$$P(A/C) = \frac{(A)/r_{CA}(C)}{(A)/r_{CA}(C) + (B)/r_{CB}(C) + 1}$$

$$P(B/C) = 1 - P(C/C) - P(A/C)$$

2. Step Two

To calculate $P(A)$, $P(B)$, and $P(C)$ from $P(A/A)\cdots P(C/C)$, it is necessary to derive at least three independent relationships involving these quantities and to then solve them simultaneously. Stoichiometric (or stationary) relationships are generally used for this purpose. Thus:

$$p(A) = P(AA) + P(BA) + P(CA)$$
$$= P(A)P(A/A) + P(B)P(A/B) + P(C)P(A/C) \tag{3}$$

$$P(B) = P(AB) + P(BB) + P(CB)$$
$$= P(A)P(B/A) + P(B)P(B/B) + P(C)P(B/C) \tag{4}$$

$$P(C) = P(AC) + P(BC) + P(CC)$$
$$= P(A)P(C/A) + P(B)P(C/B) + P(C)P(C/C) \tag{5}$$

These relationships form a set of three homogeneous equations in three unknowns. Three methods for evaluating $P(A)$, $P(B)$, and $P(C)$ from these equations will now be discussed.

   a. *Algebraic Method*. An easy to comprehend, but difficult to execute method is to simultaneously solve any two of Eqs. (3) to (5) and the normalization equation

$$P(A) + P(B) + P(C) = 1 \tag{6}$$

Thus Eqs. (4) and (6) can be used to eliminate $P(B)$ and $P(C)$ from Eq. (3). This leads to an expression for $P(A)$ in terms of conditional probabilities only; viz.,

$$P(A) =$$
$$\frac{P(A/B)\ P(B/C) - P(A/C)\ [P(B/B) - 1]}{[1-P(B/B)+P(B/C)][1+P(A/C)-P(A/A)] - [P(A/B)-P(A/C)][P(B/A)-P(B/C)]}$$
$$\tag{7}$$

$P(B)$ and $P(C)$ can be evaluated similarly. Although it is very general, the algebraic method becomes very tedious when the polymerization model is complex. Furthermore, the solutions obtained are too complex to provide any insight about polymerization mechanisms. One cannot justify the use of this method if a simpler numerical method is available.

   b. *The Method Of Price*. Price [24,67] has outlined a general method for evaluating unconditional probabilities that is simpler than the algebraic method. Variations on his approach have also been employed by others [32,34,41,59,68,72]. The homogeneous Eqs. (3) to (5) are written and manipulated in matrix form,

$$\underline{M} \cdot \underline{V} = \underline{V} \qquad\qquad (8)$$

where

$$\underline{M} = \begin{bmatrix} P(A/A) & P(A/B) & P(A/C) \\ P(B/A) & P(B/B) & P(B/C) \\ P(C/A) & P(C/B) & P(C/C) \end{bmatrix} \text{ and } \underline{V} = \begin{bmatrix} P(A) \\ P(B) \\ P(C) \end{bmatrix}$$

$\underline{M}$ is the matrix of conditional (transition) probabilities and $\underline{V}$ is the column vector of unconditional probabilities. Equation (8) can be rewritten as

$$\underline{M}' \cdot \underline{V} = 0 \qquad\qquad (9)$$

where

$$\underline{M}' = \begin{bmatrix} P(A/A) - 1 & P(A/B) & P(A/C) \\ P(B/A) & P(B/B) - 1 & P(B/C) \\ P(C/A) & P(C/B) & P(C/C) - 1 \end{bmatrix}$$

Price has shown that the elements of the vector $\underline{V}$ that represents a realistic solution of Eq. (9) are proportional to the cofactors of the elements of any row or of the diagonal of $\underline{M}'$. Evaluation of $P(A)$, $P(B)$, and $P(C)$ thus amounts to determining the cofactors, followed by normalization.

If cofactors of the elements along the diagonal of $\underline{M}'$ are evaluated and set proportional to the unconditional monomer probabilities, the following relationships are obtained:

$$P(A) = k\{[P(B/B) - 1][P(C/C) - 1] - P(B/C)P(C/B)\}$$
$$= k\{P(B/C)P(A/B) - P(A/C)[P(B/B) - 1]\} \qquad (10)$$

$$P(B) = k\{[P(A/A) - 1][P(C/C) - 1] - P(A/C)P(C/A)\}$$
$$= k\{P(B/A)P(A/C) - P(B/C)[P(A/A) - 1]\} \qquad (11)$$

$$P(C) = k\{[P(A/A) - 1][P(B/B) - 1] - P(A/B)P(B/A)\} \qquad (12)$$

The proportionality constant k, can be evaluated by substitution of Eqs. (10 to 12) into Eq. (4). This yields the following result:

$$k = \frac{1}{[1-P(B/B)+P(B/C)][1+P(A/C)-P(A/A)] - [P(A/B)-P(A/C)][P(B/A)-P(B/C)]}$$
$$\qquad (13)$$

Thus, equivalent results are obtained by the algebraic method and
by Price's approach. Although considerable mathematical manipulation is involved in the Price approach when explicit expressions
for unconditional probabilities are desired, computer routines can
be written for evaluating such quantities numerically, directly from
the matrix of conditional (transition) probabilities. This will not
be illustrated here, because there is an even simpler way to calculate the unconditional probabilities.

c. *Matrix Multiplication.* The matrix of conditional probabilities, $\underline{M}$, is a stochastic matrix since the elements of each
column total unity. Such matrices possess an extremely valuable
property for computer oriented calculations of polymer structural
features. If such matrices are multiplied by themselves repeatedly,
the elements of each row attain constant values which are equivalent
to the unconditional probabilities of interest. Normalization is
not necessary. This property is demonstrated below.

$$\underline{M} = \begin{bmatrix} P(A/A) & P(A/B) & P(A/C) \\ P(B/A) & P(B/B) & P(B/C) \\ P(C/A) & P(C/B) & P(C/C) \end{bmatrix} = \begin{bmatrix} 0.5 & 0.1 & 0.2 \\ 0.3 & 0.6 & 0.4 \\ 0.2 & 0.3 & 0.4 \end{bmatrix}$$

$$\underline{M}^2 = \underline{M} \cdot \underline{M} = \begin{bmatrix} 0.31999 & 0.16999 & 0.21999 \\ 0.40999 & 0.50999 & 0.45999 \\ 0.26999 & 0.31999 & 0.31999 \end{bmatrix}$$

$$\underline{M}^4 = \underline{M}^2 \cdot \underline{M}^2 = \begin{bmatrix} 0.23149 & 0.21149 & 0.21899 \\ 0.46449 & 0.47699 & 0.47199 \\ 0.30399 & 0.31149 & 0.30899 \end{bmatrix}$$

$$\underline{M}^8 = \underline{M}^4 \cdot \underline{M}^4 = \begin{bmatrix} 0.21841 & 0.21807 & 0.21819 \\ 0.47259 & 0.47279 & 0.47272 \\ 0.30900 & 0.30913 & 0.30908 \end{bmatrix}$$

$$\underline{M}^{16} = \underline{M}^8 \cdot \underline{M}^8 = \begin{bmatrix} 0.21818 & 0.21818 & 0.21818 \\ 0.47273 & 0.47273 & 0.47273 \\ 0.30909 & 0.30909 & 0.30909 \end{bmatrix}$$

$$\underline{M}^{32} = \underline{M}^{16} \cdot \underline{M}^{16} = \begin{bmatrix} 0.21818 & 0.21818 & 0.21818 \\ 0.47272 & 0.47272 & 0.47272 \\ 0.30909 & 0.30909 & 0.30909 \end{bmatrix}$$

The unconditional probabilities P(A), P(B), and P(C) are obtained from the elements of any column of the matrix or from the diagonal elements. The reader can verify by substituting appropriate elements of the starting matrix into Eqs. (10) to (13), that P(A) = 0.21818, P(B) = 0.47272, and P(C) = 0.30909.

Because of its simplicity and generality, we recommend that unconditional probabilities be evaluated by the matrix multiplication procedure, rather than by the other two methods described. Almost no mathematical manipulation is required by the matrix multiplication method and very little programming is required since matrix multiplication subroutines are available in most computer systems. Furthermore, the method is completely general. We have used it, for example, in a program for calculating structural aspects of terpolymers derived from a model which assumes penultimate effects for all three monomers. This required a 9 x 9 matrix of conditional probabilities.

A brief explanation of how the matrix multiplication method works is provided in Appendix A.

3. *Step Three.* Having evaluated unconditional monomer probabilities in Step two, there are no simpler structural features to calculate in the problem of terminal model terpolymerization. However, if the terpolymerization model had included penultimate effects, unconditional dyad probabilities would have been evaluated in Step two. In such a case, unconditional monomer probabilities would have been calculated as follows:

P(A) = P(AA) + P(AB) + P(AC) = P(AA) + P(BA) + P(CA)

P(B) = P(BA) + P(BB) + P(BC) = P(AB) + P(BB) + P(CB)

P(C) = P(CA) + P(CB) + P(CC) = P(AC) + P(BC) + P(CC)

In more complex models, it might be necessary to evaluate unconditional probabilities of triads, tetrads, etc. in Step two. In such situations, unconditional probabilities of smaller entities would be obtained by addition; viz.,

$$P(AB) = P(ABA) + P(ABB) + P(ABC)$$
$$= P(AAB) + P(BAB) + P(CAB)$$
$$P(AAAB) = P(AAAAB) + P(BAAAB) + P(CAAAB)$$
$$= P(AAABA) + P(AAABB) + P(AAABC)$$

4. Step Four. Having evaluated $P(A)$, $P(B)$, and $P(C)$ in Step two, dyad, triad, tetrad, and pentad distributions can now be calculated as products of these quantities and the conditional probabilities. Thus,

$$P(AA) = P(A)P(A/A)$$
$$P(AB) = P(A)P(B/A)$$
$$P(ABC) = P(A)P(B/A)P(C/B) = P(AB)P(C/B)$$
$$P(CBAC) = P(C)P(B/C)P(A/B)P(C/A)$$
$$P(ABACB) = P(A)P(B/A)P(A/B)P(C/A)P(B/C) = P(ABA)P(C/A)P(B/C)$$

These probabilities are usually not useful directly in studies of polymer behavior, so it is our practice to program the calculation of experimentally observable quantities from them. Thus, probabilities of dyads such as AB and BA are summed and reported as single entities, i.e., $P(AB + BA)$. In addition, triad and pentad probabilities are divided by monomer probabilities to obtain triad and pentad fractions. These quantities are more useful in studies of polymer spectra or reactivity than are the actual triad or pentad probabilities.

$$f_{BAB} = \frac{P(BAB)}{P(A)}$$

$$f_{(BAA + AAB)} = \frac{P(BAA) + P(AAB)}{P(A)}$$

Number and weight distributions of A, B, and C sequences can also be calculated.

$$N.D.(A_n) = P(A/A)^{n-1}[1 - P(A/A)]$$

$$N.D.(B_n) = P(B/B)^{n-1}[1 - P(B/B)]$$

$$N.D.(C_n) = P(C/C)^{n-1}[1 - P(C/C]$$

$$W.D.(A_n) = nP(A/A)^{n-1}[1 - P(A/A)]^2$$

$$W.D.(B_n) = nP(B/B)^{n-1}[1 - P(B/B)]^2$$

$$W.D.(C_n) = nP(C/C)^{n-1}[1 - P(C/C)]^2$$

It seems worthwhile to note at this point that the N.D. and W.D. functions for a terpolymerization process involving penultimate effects would have the form:

$$N.D.(A_1) = \frac{P(BA)[1 - P(A/BA)] + P(CA)[1 - P(A/CA)]}{P(BA) + P(CA)}$$

$$N.D.(A_{n>1}) = \frac{[P(BA)P(A/BA) + P(CA)P(A/CA][P(A/AA)^{n-2}][1 - P(A/AA]}{P(BA) + P(CA)}$$

$$W.D.(A_n) = \frac{n[N.D.(A_n)][P(BA) + P(CA)][1 - P(A/AA)]}{P(BA)P(A/BA) + P(CA)P(A/CA) + 1 - P(A/AA)}$$

5. *Step Five.* The steps illustrated above are adequate to calculate all structural features of terpolymers prepared according to the mechanism outlined in Step one. Step five is not needed for such a model. We illustrate the use of this step by considering next a polymerization model that is easily treated as a pseudoterpolymerization process.

C. Copolymerizations Involving Monomer Complexes

1. Steps One through Four

Consider a copolymerization process in which a complex, C, is <u>in rapid equilibrium</u> with monomers A and B and in which C adds as AB (but not as BA) to the growing chain. The following reactions govern the structure of the polymer:

$$A + B \underset{\longleftarrow}{\overset{K}{\longrightarrow}} C$$

$$\sim\sim A\cdot + A \overset{k_{AA}}{\longrightarrow} \sim\sim A\cdot$$

$$\sim\sim A\cdot \ + \ B \ \xrightarrow{\ k_{AB}\ } \ \sim\sim B\cdot$$

$$\sim\sim A\cdot \ + \ C \ \xrightarrow{\ k_{AC}\ } \ \sim\sim C\cdot \quad (= \ \sim\sim AB\cdot \ = \quad B\cdot)$$

$$\sim\sim B\cdot \ + \ A \ \xrightarrow{\ k_{BA}\ } \ \sim\sim A\cdot$$

$$\sim\sim B\cdot \ + \ B \ \xrightarrow{\ k_{BB}\ } \ \sim\sim B\cdot$$

$$\sim\sim B\cdot \ + \ C \ \xrightarrow{\ k_{BC}\ } \ \sim\sim C\cdot$$

$$\sim\sim C\cdot \ + \ A \ \xrightarrow{\ k_{CA} \ = \ k_{BA}\ } \ \sim\sim A\cdot$$

$$\sim\sim C\cdot \ + \ B \ \xrightarrow{\ k_{CB} \ = \ k_{BB}\ } \ \sim\sim B\cdot$$

$$\sim\sim C\cdot \ + \ C \ \xrightarrow{\ k_{CC} \ = \ k_{BC}\ } \ \sim\sim C\cdot$$

The equilibrium shown in the first reaction must be considered in calculating conditional probabilities because it determines the relative amounts of the polymerizable species, but it need not be considered in deriving relationships of unconditional probabilities to conditional probabilities. The other reactions are seen to be equivalent to those considered in the previous example. We can thus follow the operations of the previous example to obtain $P(A)$, $P(B)$, and $P(C)$ and unconditional probabilities for other structural aspects of the pseudo-terpolymer.

2. Step Five

In this step, unconditional probabilities for the actual AB copolymer are derived from unconditional probabilities calculated in the previous four steps for the pseudo-terpolymer. This is accomplished by adding unconditional probabilities for the pseudo-terpolymer followed by renormalization. Renormalization is necessary because each C unit in the pseudo-terpolymer represents two units (i.e., AB) in the actual copolymer. If $P'(\ )$ and $P(\ )$ represent unconditional probabilities for the actual copolymer and the pseudo-terpolymer, respectively, the following relationships are easily derived.

$$P'(A) \propto P(A) + P(C)$$

$$P'(B) \propto P(B) + P(C)$$

$$P'(A) = \frac{P(A) + P(C)}{P(A) + P(B) + P(C) + P(C)}$$

$$P'(A) = \frac{P(A) + P(C)}{1 + P(C)}$$

$$P'(B) = \frac{P(B) + P(C)}{1 + P(C)}$$

$$P'(AA) = \frac{P(AA) + P(AC)}{1 + P(C)}$$

$$P'(AB) = \frac{P(AB) + P(C)}{1 + P(C)}$$

$$P'(BA) = \frac{P(CA) + P(BA) + P(BC) + P(CC)}{1 + P(C)}$$

$$P'(BB) = \frac{P(CB) + P(BB)}{1 + P(C)}$$

$$P'(AAA) = \frac{P(AAA) + P(AAC)}{1 + P(C)}$$

$$P'(BAA) = \frac{P(BAA) + P(CAA) + P(BAC) + P(CAC)}{1 + P(C)}$$

$$P'(AAB) = \frac{P(AAB) + P(AC)}{1 + P(C)}$$

$$P'(BAB) = \frac{P(BAB) + P(BC) + P(CC) + P(CAB)}{1 + P(C)}$$

$$f(BAB) = \frac{P'(BAB)}{P'(A)}$$

The calculation of tetrad distributions, pentad distributions, pentad fractions, etc. proceeds similarly. The normalization factor, $1/[1 + P(C)]$, applies in all cases.

Seiner and Litt [71,74] used a steady-state kinetic approach to derive expressions for calculating triad distributions of copolymers derived from systems involving monomer complexes. To simplify the mathematics, they assumed that one of the reactivity ratios was zero. Such simplification is unnecessary in the computer oriented method outlined here. In fact it can be easily

extended to more complex models and to the calculation of other
structural features such as pentad fractions, etc. We have used
this approach to calculate structural features of styrene-methyl
methacrylate copolymers derived from styrene-methacrylic anhydride
copolymers, for example [79].

D. Reversible Copolymerization

The method outlined to this point is very general and is applicable to a wide variety of polymerization models, including
multicomponent polymerizations, penultimate and penpenultimate
effects, stereosequence effects, monomer complexation and multi-
state propagation species. In all these situations the propagation
process proceeds forward in a series of discrete steps. It is
therefore possible to calculate conditional probabilities <u>directly</u>
from kinetic parameters and monomer concentrations. When the polymerization processes are <u>reversible,</u> the conditional probabilities
cannot be calculated directly; they must be evaluated simultaneously with the unconditional probabilities. O'Driscoll and co-workers
[63,77] have described how to set up equations for such problems,
but there does not seem to be any convenient way to solve them.
Approximation methods are used, for which the computer is a definite
asset, if not a necessity. The general approach outlined in this
chapter is not presently applicable to reversible polymerization
systems. The reader is therefore referrred to O'Driscoll's papers
for instruction in the analysis of such systems. O'Driscoll provides a Monte Carlo program for reversible terpolymerization in this
volume.

V. CALCULATIONS FOR POLYMERS PREPARED IN HIGH CONVERSION

The methods described to this point are adequate for polymers
prepared under conditions where the ratios of monomers in the polymerization mixture remain constant. Such is the case for "azeotropic" polymers or polymers prepared in low conversion. To obtain
structural features of polymers prepared in high conversion, it is

necessary to calculate average values from instantaneous values calculated at the beginning and end of various conversion intervals,
the consumption of monomers during each interval having been taken
into account. Ito and Yamashita [80,81] have shown how this can be
done graphically, but the task is much easier if done by numerical
integration with the computer [38,49,50].

The average distribution of AA dyads in a high conversion copolymer, for example, can be calculated by either Eq. (14) (Fig. 1)
or Eq. (15) (Fig. 2), where P(AA) i represents the distribution of
AA dyads for the polymer formed at the start of a conversion increment (mole percent conversion), and n is the number of increments
required to obtain the final conversion.

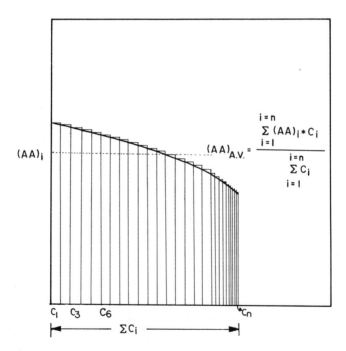

Fig. 1. Calculation of the average AA dyad distribution by
use of a step integration function [Eq. (14)].

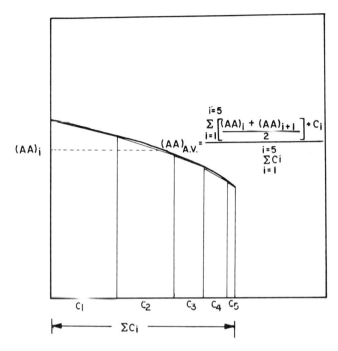

FIG. 2. Calculation of the average AA dyad distribution by use of a trapezoidal integration function [Eq. (15)].

$$P(AA)_{AV.} = \frac{\displaystyle\sum_{i=1}^{i=n} [P(AA)_i * C_i]}{\displaystyle\sum_{i=1}^{i=n} C_i} \qquad (14)$$

$$P(AA)_{AV.} = \frac{\displaystyle\sum_{i=1}^{i=n} \left[\frac{P(AA)_i + P(AA)_{i+1}}{2}\right] * C_i}{\displaystyle\sum_{i=1}^{i=n} C_i} \qquad (15)$$

Similarly, the average BAB triad fraction for a high conversion
polymer could be calculated from either Eq. (16) or (17).

$$f_{BAB(AV.)} = \frac{\sum_{i=1}^{i=n} [P(BAB)_i * C_i]}{\sum_{i=1}^{i=n} [P(A)_i * C_i]} \tag{16}$$

$$f_{BAB(AV.)} = \frac{\sum_{i=1}^{i=n} \left[\frac{P(BAB)_i + P(BAB)_{i+1}}{2}\right] * C_i}{\sum_{i=1}^{i=n} \frac{[P(A)_i + P(A)_{i+1}]}{2} * C_i} \tag{17}$$

Programming this part of a calculation requires very few statements
if the quantities of interest are manipulated as subscripted variables, as is seen in Appendix B. The calculation time can be reduced if summations for the largest structural features (e.g., pentads) of interest are calculated and if summations for the other
features are calculated from them, using the stoichiometric
(stationary) relationships previously discussed; viz.,

$$\sum_{i=1}^{i=n} \frac{[P(A)_i + P(A)_{i+1}]}{2} * C_i = \sum_{i=1}^{i=n} \frac{[P(AA)_i + P(AA)_{i+1}]}{2} * C_i$$

$$+ \sum_{i=1}^{i=n} \frac{[PAB)_i + P(AB)_{i+1}]}{2} * C_i$$

VI. SUMMARY

By following the general steps outlined in this chapter, it is
possible to calculate structural features of multicomponent polymers
formed by a variety of mechanisms in high as well as low conversion.

If the matrix multiplication technique is used to obtain unconditional probabilities (stationary concentrations), the results can
be obtained almost without algebraic manipulation.

To serve as a model for programming such calculations, a program written to calculate several structural features of terpolymers formed by a terminal model process is included in Appendix B.
The program utilizes the matrix multiplication method to evaluate
the mole fractions of monomers present and it uses a trapezoidal
function for numerical integration.

The following programs, which are more extensive than the
sample program provided, are available from the authors.

SEQDA [50]: Calculates the compositions, dyad distributions,
triad fractions, tetrad distributions, pentad fractions, number and
weight distributions of monomer sequences for copolymers prepared
in either low or high conversion by terminal or penultimate model
processes.

SEQDB [79]: Calculates the compositions, triad fractions and
pentad fractions for methyl methacrylate containing copolymers derived from methacrylic anhydride containing copolymers. This program is useful for studies on cyclocopolymerization.

SEQDC: Calculates the compositions, dyad distributions, triad
fractions, tetrad distributions, B-centered pentad fractions, number and weight distributions of monomer sequences for terpolymers
prepared in either low or high conversion from terminal or penultimate model processes.

SEQDD: Calculates structural features for copolymers formed
from copolymerizations in which monomer complexes participate in
propagation processes.

APPENDIX A: Basis for the Matrix Multiplication Method

The matrices of conditional probabilities employed in Step two
of the general procedure correspond to transition probability matrices in the theory of Markov chains [82,83]. Such matrices can
be used to predict probabilities of the states of a system after a

given number of events has occurred, provided that the system is
described at the start. If unconditional probabilities of finding
the system in state i after n events (n = 0, 1, 2, ...) are written
as a column vector $\underline{V}(n) = [V_1(n), V_2(n), V_3(n), ...]$, and if the
transition matrix is represented by $\underline{M}$, where each element of the
matrix, m_{ij}, represents the conditional probability $P.(i/j)$, the
following relationships hold.

$$\underline{V}(1) = \underline{M} \cdot \underline{V}(0)$$
$$\underline{V}(2) = \underline{M} \cdot \underline{V}(1) = \underline{M} \cdot [\underline{M} \cdot \underline{V}(0)] = \underline{M}^2 \cdot \underline{V}(0)$$
$$\underline{V}(3) = \underline{M} \cdot \underline{V}(2) = \underline{M} \cdot [\underline{M} \cdot \underline{V}(1)] = \underline{M}^3 \cdot \underline{V}(0)$$
$$\underline{V}(n) = \underline{M}^n \cdot \underline{V}(0)$$

After a sufficient number of events have occurred, the state of the
system becomes independent of its initial state. (This is equivalent to saying that the polymerization process becomes independent
of the initiation step.) Thus, $\underline{V}(n)$ and $\underline{M}^n$ approach limiting forms,
$\underline{V}(\infty)$ and $\underline{M}^\infty$, as n increases. The elements in each row of $\underline{M}^\infty$ are
identical and those in each column are identical to corresponding
elements in $\underline{V}(\infty)$. $\underline{M}^\infty$ has such a structure that $\underline{V}(\infty)$ is obtained
when $\underline{M}^\infty$ is multiplied by any vector, including $\underline{V}(\infty)$. $\underline{V}(\infty)$ thus
contains unconditional probabilities of the states of the system
after stationarity has been attained. One could also calculate
$\underline{V}(\infty)$ by starting with any $\underline{V}(o)$ and continuously multiplying the
transition probability matrix by $\underline{V}(0)$, $\underline{V}(1)$, ..., $\underline{V}(n)$, etc., until
a limit is obtained. Convergence is faster when $\underline{M}^n$ is repeatedly
squared, however.

APPENDIX B: AN INSTRUCTIVE MODEL PROGRAM

```
C  THIS PROGRAM IS WRITTEN TO SERVE AS A MODEL FOR READERS
C  INTERESTED IN PROGRAMMING SEQUENCE DISTRIBUTION CALCULATIONS.
C  THE PROGRAM IS DESIGNED TO CALCULATE REPRESENTATIVE ASPECTS
C  OF THE STRUCTURE OF TERPOLYMERS DERIVED FROM TERMINAL MODEL
C  (1ST ORDER MARKOVIAN) POLYMERIZATION PROCESSES. MORE COMPLETE
C  PROGRAMS APPLICABLE ALSO TO 2ND ORDER MARKOVIAN PROCESSES ARE
C  AVAILABLE FROM THE AUTHORS.
      DIMENSION P(3,3),X(3,3),Y(3,3),D(9),T(9),AND(9),AWD(9),
     1 FD(9),FT(9),FAND(9),FAWD(9),TD(9),TT(9),TAND(9),TAWD(9),N(9)
C  INITIALIZE: AREAS FOR TOTALING A(TA),B(TB),C(TC),MOLAR CONVERSION
C  (TCM),DYADS(TD),TRIADS(TT),NUMBER DISTRIBUTIONS(TAND),AND
C  WEIGHT DISTRIBUTIONS(TAWD) OF A-SEQUENCES ARE SET EQUAL TO ZERO.
      DATA TA,TB,TC,TCM/0.,0.,0.,0./,TD,TT,TAND,TAWD/36*0. /,FIRSTA/-0.5/
C  INPUT: MONOMER REACTIVITY RATIOS RAB,RBA,RAC,RCA,RBC,RCB. RAB MEANS
C  THE RATIO OF RATE CONSTANTS FOR RADICAL A TC MONOMERS A AND B, I.E.,
C  KAA/KAB. AF AND BF ARE THE MOLAR PERCENTAGES OF MONOMERS A AND B
C  INITALLY PRESENT IN THE POLYMERIZATION MIXTURE.CON/ IS THE CONVERSION
C  IN WEIGHT PERCENT. AMW,BMW, AND CMW ARE MONOMER MO_ECULAR WEIGHTS.
      READ(5,1) RAB,RBA
    1 FORMAT(3F10.5)
      READ(5,1) RAC,RCA
      READ(5,1) RBC,RCB
      READ(5,1) AF,BF,CONV
      READ(5,1) AMW,BMW,CMW
      CF = 100. - AF - BF
      WEIGHT = AF*AMW + BF*BMW + CF*CMW
```

```
C     CONVERSION INCREMENT (CI) FOR NUMERICAL INTEGRATION IS INITALLY
C     0.005 (0.5 MOLE PERCENT). THE TRAPEZOIDAL RULE IS FOLLOWED.
      CI = 0.005
      CIX = CI
      WRITE(6,2)RAB,RBA,RAC,RCA,RBC,RCB ,AF,BF,CF,AMW,BMW,CMW
C     CALCULATION OF CONDITIONAL PROBABILITIES - P(I,J) IS CONSIDERED
C     AS P(I/J), ETC., 1=A,2 = B,3 = C.THE INEFFICIENT PROGRAMMING HERE
C     IS DESIGNED FOR THE READER , NOT THE COMPUTER.
   31 P(1,1) = 1./(1. + BF/(RAB*AF) + CF/(RAC*AF))
      P(2,1) = (BF/(RAB*AF))/(1. + BF/(RAB*AF) + CF/(RAC*AF))
      P(3,1) = 1. - P(1,1) - P(2,1)
      P(1,2) = (AF/(RBA*BF))/(AF/(RBA*BF) + 1. + CF/(RBC*BF))
      P(2,2) = 1./(AF/(RBA*BF) + 1. + CF/(RBC*BF))
      P(3,2) = 1. - P(1,2) - P(2,2)
      P(1,3) = (AF/(RCA*CF) )/(AF/(RCA*CF) + BF/(RCB*CF) + 1.)
      P(3,3) = 1./(AF/(RCA*CF) + BF/(RCB*CF) + 1.)
      P(2,3) = 1. - P(1,3) - P(3,3)
C     MATRIX MULTIPLICATION STEP - GMPRD IS AN IBM FORTRAN SUBROUTINE FOR
C     MULTIPLYING MATRICES - SEE THE IBM SCIENTIFIC SUBROUTINE PACKAGE
C     FOR INSTRUCTIONS IN ITS USE - 3X3 MATRIX A IS MULTIPLIED BY 3X3
C     MATRIX B AND THE RESULT IS STORED IN 3X3 MATRIX C WHEN THE SUBROUTINE
C     IS CALLED AS FOLLOWS:  CALL GMPRD(A,B,C,3,3,3)
      CALL GMPRD(P,P,X,3,3,3)
    3 CALL GMPRD(X,X,Y,3,3,3)
      CALL GMPRD(Y,Y,X,3,3,3)
      DO 4 I = 1,3
      DO 4 J = 1,3
      JJ = J + 1
      IF(ABS(X(I,J)-X(I,JJ)).GT.0.00001) GO TO 3
    4 CONTINUE
```

```
C  CALCULATION OF MOLE FRACTIONS (UNCONDITIONAL PROBABILIT-ES) OF
C  MONOMERS IN THE TERPOLYMER.THESE WILL BE DESIGNATED A,B, AND C.
      A = X(1,1)
      B = X(2,2)
      C = X(3,3)
C  CALCULATION OF DYAD DISTRIBUTIONS:1=AA,2=BA,3=CA,4=AB,5=BB,6=CB,
C  7=AC,8=BC,9=CC.
      D(1) = A * P(1,1)
      D(2) = B * P(1,2)
      D(3) = C * P(1,3)
      D(4) = A * P(2,1)
      D(5) = B * P(2,2)
      D(6) = C * P(2,3)
      D(7) = A * P(3,1)
      D(8) = B * P(3,2)
      D(9) = C * P(3,3)
C  CALCULATION OF A-CENTERED TRIAD DISTRIBUTIONS - 1=AAA,2=AAB,3=AAC,
C  4=BAA,5=BAB,6=BAC,7=CAA,8=CAB,9=CAC.B- AND C- CENTERED TRIAD
C  DISTRIBUTIONS ARE CALCULATED SIMILARLY.
      T(1) = D(1)*P(1,1)
      T(2) = D(1) * P(2,1)
      T(3) = D(1) *P(3,1)
      T(4) = D(2) * P(1,1)
      T(5) = D(2) * P(2,1)
      T(6) = D(2) * P(3,1)
      T(7) = D(3) * P(1,1)
      T(8) = D(3) * P(2,1)
      T(9) = D(3) * P(3,1)
C  CALCULATION OF NUMBER(AND) AND WEIGHT(AWD) DISTRIBUTIONS OF A-
```

```
C  SEQUENCES. SIMILAR PROGRAMMING APPLIES FOR THE B- AND C- SEQUENCES.
      DO 6 I = 1,9
      XI = I
      AND(I) = (P(1,1)**(I-1))*(1.-P(1,1))
    6 AWD(I) = AND(I)*XI*(1. - P(1,1))
C  START NUMERICAL INTEGRATION - TEST FOR FIRST CALCULATION.
      IF(FIRSTA.LT.0.) GO TO 10
C  TOTAL THE STRUCTURAL FEATURES INCORPORATED IN TERPOLYMER DURING
C  CONVERSION INCREMENT.
C  XX IS USED TO AVOID REPEATEDLY CALCULATING CI/2.
   16 XX = CI/2.
      TA = (A + FIRSTA)*XX + TA
      TB = (B + FIRSTB) * XX + TB
      TC = (C + FIRSTC) * XX + TC
      TCM = TCM + CI
      DO 7 I = 1,9
      TD(I) = TD(I) + (FD(I) + D(I))*XX
      TT(I) = TT(I) + (FT(I) + T(I))*XX
      TAND(I) = TAND(I) + (FAND(I) + AND(I))*XX
    7 TAWD(I) = TAWD(I) + (FAWD(I) + AWD(I))*XX
C  TEST FOR FINAL CONVERSION.CW = TOTAL CONVERSION IN WEIGHT PERCENT.
      CW = 10000.*(TA*AMW + TB*BMW + TC*CMW)/WEIGHT
      IT = (CONV - CW)*1000.
      IF (IT) 12,13,10
C  SET VALUES AT START OF INCREMENT.
   10 FIRSTA = A
      FIRSTB = B
```

```
      FIRSTC = C
      DO 11 I = 1,9
      FD(I) = D(I)
      FT(I) = T(I)
      FAND(I) = AND(I)
   11 FAWD(I) = AWD(I)
C ADJUST MONOMER CONTENTS FOR CONSUMPTION DURING INCREMENT AND
C REDUCE CONVERSION INCREMENT IF NECESSARY.THE POSSIBILITY OF
C MONOMER DEPLETION IS ALSO TESTED HERE.
      AFEST = AF - A*CI*100.
      BFEST = BF - B*CI*100.
      CFEST = CF - C*CI*100.
      IF(AFEST.LT.0.000001) GO TO 20
      IF(BFEST.LT.0.000001) GO TO 20
      IF(CFEST.LT.0.000001) GO TO 20
      AF = AFEST
      BF = BFEST
      CF = CFEST
C NOW CALCULATE VALUES AT END OF THE CONVERSION INCREMENT.
      GO TO 31
   20 CI = CI*0.1
      IJK = 10000. - CIX/CI
      IF(IJK.GT.0) GO TO 10
      WRITE(6,8)
    8 FORMAT(2X,'MONOMER DEPLETION')
      GO TO 13
```

```
C     CI IS TOO LARGE BECAUSE SPECIFIED CONVERSION WAS EXCEEDED.
C     RESET TOTALS TO VALUES PREVAILING AT START OF INCREMENT AND
C     REDUCE CI.
12    TA = TA - (A + FIRSTA)*XX
      TB = TB - (B + FIRSTB)*XX
      TC = TC - (C + FIRSTC)*XX
      TCM = TCM - CI
      DO 15 I = 1,9
      TD(I) = TD(I) - (FD(I) + D(I))*XX
      TT(I) = TT(I) - (FT(I) = T(I))*XX
      TAND(I) = TAND(I) - (FAND(I) + AND(I))*XX
15    TAWD(I) = TAWD(I) - (FAWD(I) + AWD(I))*XX
C     REDUCE CONVERSION INCREMENT.
      CI = CI * 0.1
      GO TO 16
C     BEGIN OUTPUT: CALCULATE AVERAGE VALUES,TRIAD FRACTIONS.
13    A = (TA/TCM)*100.
      B = (TB/TCM)*100.
      C = (TC/TCM)*100.
      DO 30 I = 1,9
      N(I) = I
      D(I) = TD(I)/TCM
      T(I) = TT(I)/TA
      AND(I) = TAND(I)/TCM
30    AWD(I) = TAWD(I)/TCM
C     OBTAIN SUMS OF STRUCTURAL FEATURES THAT MAY NOT BE DISTINGUISHABLE
```

```
C    EXPERIMENTALLY.
     AB = D(2) + D(4)
     AC = D(3) + D(7)
     BC = D(6) + D(8)
     AAB = T(2) + T(4)
     AAC = T(3) + T(7)
     BAC = T(6) + T(8)
     WRITE (6,5)CW,A,B,C,(D(I),I=1,9),AB,AC,BC,(T(I),I=1,9),
    1AAB,AAC,BAC,(N(I),AND(I),I=1,9)
   5 FORMAT(3X,'CONVERSION=',F10.5,'WEIGHT PERCENT'//3X,
    1'TERPOLYMER COMPOSITION'/3X,'A =',F10.5,3X,'B =',F10.5,3X,
    2'C =',F10.5//3X,'DYAD DISTRIBUTIONS'/3X,'AA =',F10.5,3X,'BA =',
    3F10.5,3X,'CA =',F10.5,/3X,'AB =',F10.5,3X,'BB =',F10.5,3X,
    4'CB =',F10.5,/3X,'AC =',F10.5,3X,'BC =',F10.5,3X,'CC =',F10.5,
    5/3X,'AB + BA =',F10.5,3X,'AC + CA =',F10.5,3X,'BC + CB =',
    6F10.5//3X,'A-CENTERED TRIAD FRACTIONS'/3X,'AAA =',F10.5,
    73X,'AAB =',F10.5,3X,'AAC =',F10.5,3X,'BAA =',F10.5,3X,'BAB =',
    8F10.5,3X,'BAC =',F10.5,3X,'CAA =',F10.5,3X,'CAB =',F10.5,
    93X,'CAC =',F10.5,3X,'AAB + BAA + CAA =',F10.5,3X,'AAC + CAA =',F10.5,
    03X,'BAC + CAB =',F10.5//3X,'NUMBER(N.D.) AND WEIGHT(W.D.)',
    1'DISTRIBUTIONS OF A-SEQUENCES'/2X,'N',6X,'N.D.',9X,'W.D.'/
    29(2X,I1,F10.5,3X,F10.5/))
   2 FORMAT(3X,'MONOMER REACTIVITY RATIOS'/3X,'RAB =',F10.5,3X,
    1'RBA =',F10.5/3X,'RAC =',F10.5,3X,'RCA =',F10.5/3X,'RBC =',
    2F10.5,3X,'RCB =',F10.5/3X,'MONOMER FEED COMPOSITION'/3X,
    3'A =',F10.5,3X,'B =',F10.5,3X,'C =',F10.5//3X,'MONOMER',
    4'MOLECULAR WEIGHTS'/3X,'A =',F10.5,3X,'B =',F10.5,3X,'C =',
    5F10.5/)
     STOP
     END
   *   TYPICAL   OUTPUT   *
```

MONOMER REACTIVITY RATIOS
RAB = 0.30000   RBA = 0.70000
RAC = 0.50000   RCA = 0.50000
RBC = 0.10000   RCB = 0.10000

MONOMER FEED COMPOSITION
A = 50.00000   B = 25.00000   C = 25.00000

MONOMER MOLECULAR WEIGHTS
A = 100.00000   B = 100.00000   C = 100.00000

CONVERSION = 4.99983 WEIGHT PERCENT

TERPOLYMER COMPOSITION
A = 24.98708   B = 38.30408   C = 36.70554

DYAD DISTRIBUTIONS
AA = 0.06947   BA = 0.08058   CA = 0.09982
AB = 0.11268   BB = 0.02745   CB = 0.24291
AC = 0.06772   BC = 0.27500   CC = 0.02433
AB + BA = 0.19326   AC + CA = 0.16754   BC + CB = 0.51791

A-CENTERED TRIAD FRACTIONS
AAA = 0.07731   AAB = 0.12537   AAC = 0.07535
BAA = 0.08966   BAB = 0.14543   BAC = 0.08741
CAA = 0.11106   CAB = 0.18014   CAC = 0.10827
AAB + BAA = 0.21503   AAC + CAA = 0.18641   BAC + CAB = 0.26755

NUMBER(N.D.) AND WEIGHT(W.D.) DISTRIBUTIONS OF A-SEQUENCES

N	N.D.	W.D.
1	0.72201	0.52131
2	0.20070	0.28981
3	0.05580	0.12085
4	0.01551	0.04480
5	0.00431	0.01557
6	0.00120	0.00520
7	0.00033	0.00169
8	0.00009	0.00054
9	0.00003	0.00017

REFERENCES

1. T. Alfrey and G. Goldfinger, J. Chem. Phys., 12, 205, 244, 322 (1944).

2. F. T. Wall, J. Am. Chem. Soc., 66, 2050 (1944).

3. F. R. Mayo and F. M. Lewis, J. Am. Chem. Soc., 66, 1594 (1944).

4. C. Walling and E. R. Briggs, J. Am. Chem. Soc., 67, 1774 (1945).

5. E. Merz, T. Alfrey, and G. Goldfinger, J. Polymer Sci., 1, 75 (1946).

6. H. L. Frisch, C. Schuerch, and M. Szwarc, J. Polymer Sci., 11, 559 (1953).

7. B. D. Coleman, J. Polymer Sci., 31, 155 (1958).

8. G. Ham, J. Polymer Sci., 31, 155 (1958).

9. T. Alfrey and A. V. Tobolsky, J. Polymer Sci., 38, 269 (1959).

10. C. Schuerch, J. Polymer Sci., 40, 533 (1959).

11. J. W. L. Fordham, J. Polymer Sci., 39, 321 (1959).

12. F. A. Bovey and G. V. D. Tiers, J. Polymer Sci., 44, 173 (1960).

13. G. E. Ham, J. Polymer Sci., 45, 177 (1960).

14. G. G. Lowry, J. Polymer Sci., 42, 463 (1960).

15. R. L. Miller and L. E. Nielsen, J. Polymer Sci., 46, 303 (1960).

16. A. Miyake and R. Chujo, J. Polymer Sci., 46, 163 (1960).

17. G. Natta, G. Mazzanati, A. Valvassori, G. Sartori, and D. Morero, Chim. and Ind., (Milano), 42, 125 (1960).

18. S. Newman, J. Polymer Sci., 47, 111 (1960).

19. U. Johnsen, Kolloid Z., 178, 161 (1961).

20. R. L. Miller, J. Polymer Sci., 56, 375 (1962).

21. R. L. Miller, J. Polymer Sci., 57, 975 (1962).

22. L. Peller, J. Phys. Chem., 66, 685 (1962).

23. L. Peller, J. Chem. Phys., 36, 2976 (1962).

24. F. P. Price, J. Chem. Phys., 36, 209 (1962).

25. B. D. Coleman and T. G. Fox, J. Chem. Phys., 38, 1065 (1963).

26. B. D. Coleman and T. G. Fox, J. Polymer Sci., A1, 3183 (1963).

27. H. K. Frensdorf and R. Pariser, J. Chem. Phys., 39, 2303 (1963).

28. J. Hijmans, Physica, 29, 1, 819 (1963).

29. S. Igashari, *J. Polymer Sci.*, Part B, *1*, 359 (1963).

30. W. Ring, *J. Polymer Sci.*, Part B, *1*, 323 (1963).

31. R. Chujo, S. Satoh, and E. Nagai, *J. Polymer Sci.*, Part A, *2*, 895 (1964).

32. T. Fueno and J. Furukawa, *J. Polymer Sci.*, *A2*, 3681 (1964).

33. H. J. Harwood and W. M. Ritchey, *J. Polymer Sci.*, Part B, *2*, 601 (1964).

34. T. Fueno, R. A. Shelden, and J. Furukawa, *J. Polymer Sci.*, Part A, *3*, 1279 (1965).

35. K. Ito and Y. Yamashita, *J. Polymer Sci.*, Part A, *3*, 2165 (1965).

36. J. B. Kinsinger and D. Colton, *J. Polymer Sci.*, Part B, *3*, 797 (1965).

37. L. Peller, *J. Chem. Phys.*, *43*, 2355 (1965).

38. J. A. Seiner, *J. Polymer Sci.*, Part A, *3*, 2401 (1965).

39. R. Chujo, *J. Phys. Soc.*, (Japan), *21*, 2669 (1966).

40. B. D. Coleman, T. G. Fox, and M. Reinmoller, *J. Polymer Sci.*, Part B, *4*, 1029 (1966).

41. H. L. Frisch, C. L. Mallows, and F. A. Bovey, *J. Chem. Phys.*, *45*, 1565 (1966).

42. U. Johnsen, *Ber. Bunsen Ges. fur Physik. Chem.*, *70*, 320 (1966).

43. M. Reinmoller and T. G. Fox, *A.C.S. Polymer Preprints*, *7*, 987, 999 (1966).

44. R. Chujo, *Makromol. Chem.*, *107*, 142 (1967).

45. W. Ring, *Makromol. Chem.*, *101*, 145 (1967).

46. C. Tosi, *Makromol. Chem.*, *108*, 307 (1967).

47. A. Valvassori and G. Sartori, *Adv. Polymer Sci.*, *5*, 28 (1967).

48. R. Chujo, *J. Macromol. Sci.-Physics*, *B2*, 1 (1968).

49. H. J. Harwood, N. W. Johnston, and H. Piotrowski, *J. Polymer Sci.*, Part C, *25*, 23 (1968).

50. H. J. Harwood, *J. Polymer Sci.*, Part C, *25*, 37 (1968).

51. W. Kawai, *J. Polymer Sci.*, Part A1, *6*, 1945 (1968).

52. L. D. Maxim, C. H. Kuist, and M. E. Meyer, *Macromolecules*, *1*, 86 (1968).

53. J. Schaefer, *Macromolecules*, *1*, 111 (1968).

54. C. Tosi, *Adv. Polymer Sci.*, *5*, 451 (1968).

55. E. Klesper and W. Gronski, *J. Polymer Sci.*, Part B, *7*, 661 (1969).

56. P. L. Luisi and R. M. Mazo, *J. Polymer Sci.*, Part A2, 7, 775 (1969).

57. R. A. Shelden, T. Fueno, and J. Furukawa, *J. Polymer Sci.*, Part A2, 2, 763 (1969).

58. M. H. Theil, *Macromolecules*, 2, 137 (1969).

59. C. Tosi and G. Allegra, *Makromol. Chem.*, 129, 275 (1969).

60. S. Yabumoto, K. Ishii, and K. Arita, *J. Polymer Sci.*, Part A-1, 7, 1577 (1969).

61. A. Blumstein, S. L. Malkotra, and A. C. Watterson, *J. Polymer Sci.*, Part A2, 8, 1599 (1970).

62. D. R. Cruse and R. G. Lacombe, *J. Polymer Sci.*, Part A1, 8, 1373 (1970).

63. J. A. Howell, M. Izu, and K. F. O'Driscoll, *J. Polymer Sci.*, Part A1, 8, 699 (1970).

64. M. Izu and K. F. O'Driscoll, *Polymer J.*, (Japan), 1, 27 (1970).

65. M. Izu and K. F. O'Driscoll, *J. App. Polymer Sci.*, 14, 1515 (1970).

66. E. Klesper, *J. Polymer Sci.*, Part A1, 8, 1191 (1970).

67. F. P. Price, in "Markov Chains and Monte Carlo Calculations in Polymer Science", (G. G. Lowry, ed.), Marcel Dekker, Inc., New York, 1970, Chap. 7.

68. C. W. Pyun, *J. Polymer Sci.*, Part A2, 8, 1111 (1970).

69. G. V. Strate, *J. Appl. Polymer Sci.*, 14, 2509 (1970).

70. H. K. Frensdorf, *Macromolecules*, 4, 369 (1971).

71. M. Litt and J. A. Seiner, *Macromolecules*, 4, 314 (1971).

72. C. W. Pyun, *J. Polymer Sci.*, Part A2, 9, 577 (1971).

73. C. W. Pyun and T. G. Fox, *J. Polymer Sci.*, Part A2, 9, 615 (1971).

74. J. A. Seiner and M. Litt, *Macromolecules*, 4, 308 (1971).

75. E. Tsuchida and T. Tomoro, *Makromol. Chem.*, 141, 265 (1971).

76. R. Chujo, M. Kamei, and A. Nishioka, *Polymer J. (Japan)*, 3, 289 (1972).

77. B. Kang, K. F. O'Driscoll, and J. A. Howell, *J. Polymer Sci.*, Part A-1, 10, 2349 (1972).

78. W. Gronski, E. Klesper, and H-J. Cantow, *J. Polymer Sci.*, Part C, 42, 217 (1973).

79. D. L. Newman, Ph.D. Dissertation, The University of Akron, 1971.

80. Y. Yamashita and K. Ito, *J. App. Polymer Sci.*, App. Polymer Symposia, *8*, 245 (1969).

81. Y. Yamashita, K. Ito, H. Ishii, S. Hoshino, and M. Kai, *Macromolecules, 1*, 529 (1968).

82. J. M. Myhre, "Markov Chains and Monte Carlo Calculations in Polymer Science", (G. G. Lowry, ed.), Marcel Dekker, Inc., New Yori, 1970, Chap. 2.

83. J. Kemeny and J. L. Snell, "Finite Markov Chains", Van Nostrand Publishing Co., Princeton, 1960.

Chapter 3

SIMULATION OF POLYMERIZATION
AT THE MICROSCOPIC LEVEL

Kenneth F. O'Driscoll

Department of Chemical Engineering
University of Waterloo
Waterloo, Ontario

I. INTRODUCTION

In a polymerization reaction, the polymeric product consists
of a collection of macromolecules which contain information about
the particular reactions which occurred in the course of polymerization. This information is present in the form of such macromolecular parameters as molecular weight distribution, stereoregularity, head to head or head to tail structures, and, in the case of
copolymers, composition distribution and sequence distribution.
From experimental analyses of such information it is often possible
to deduce a kinetic scheme which accurately describes the polymerization reactions. Conversely, given sufficient information about
a polymerizing system it is reasonable to suppose that one could
work in the other direction and that simulation could be a valuable
tool to predict polymer properties or to probe some kinetic or
mechanistic aspects of the polymerization process. The results of
such simulations have proved useful for discerning unsuspected
things about complex kinetic models. Simulation in polymerization
has also been used at the macroscopic level in reactor design. However, this chapter will be restricted to considerations at the molecular or microscopic level.

The technique which has been used to simulate polymerization
is a straightforward application of the Monte-Carlo method. In
this technique, probabilities of various reaction events occurring
are computed after each reaction. These probabilities are then
each assigned to a proportionate space in a number range from 0 to
1. A random number generator is used to produce a number between 0
and 1 with any desired precision, and this number then determines
the next event which occurs. If this reaction event changes the
probabilities in any way, a calculation may then be performed to
determine the probabilities for the next event. Perhaps the best
way to illustrate the technique is with an example taken from
Lowry's book [1].

Consider a polymer chain of length n which can either propa-
gate with probability p, depropagate with probability d, or terminate with probability t.

$$M_n^* + M \underset{d}{\overset{p}{\rightleftharpoons}} M_{n+1}^*$$

$$M_n^* \xrightarrow{t} \text{Inactive polymer}$$

$$p + d + t = 1$$

A flow chart for the logic in the computer simulation of this reaction sequence is given in Fig. 1. The only input data required are the normalized probabilities p and d. This program was designed to produce a total of 1000 molecules of variable chain lengths whose molecular weight distribution could be analyzed. However, with only minor changes in the logic and input requirements, this type of program can handle many different models of polymerization processes. The following sections will review a variety of polymerization processes which have been simulated.

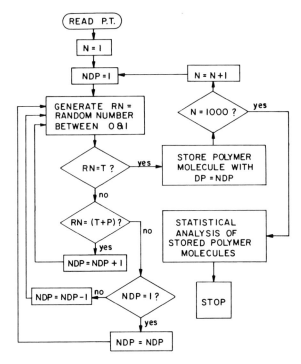

FIG. 1. Flow chart for simulation of reversible homopolymerization [1].

II. COPOLYMERIZATION

A. Composition

In the simplest model of binary copolymerization a growing
polymer chain, ending in monomer M_i, adds on another monomer unit
M_j, which may or may not be the same as M_i.

$$M_i^* + M_j \longrightarrow M_iM_j^* \qquad (i, j = 1, 2)$$

The rate of this reaction is characterized by a second-order rate
constant k_{ij}, and experimental data on copolymer composition as
a function of monomer feed composition yield reactivity ratios,
$r_{ij} = k_{ii}/k_{ij}$. The conditional probability of a given chain end
M_i adding a particular monomer M_j is given by

$$P_{ij} = \frac{1}{1 + r_{ij}x_i/(1 - x_i)}$$

where x_i is the mole fraction of M_i in the monomer feed. Because
it is a binary system,

$$P_{ii} + P_{ij} = 1$$

For systems of three or more monomers, the above equations may be
generalized. For example, in a three-component system, the conditional probabilities for a chain ending in M_1 are

$$P_{11} + P_{12} + P_{13} = 1$$

and

$$P_{12} = \frac{1}{1 + r_{12}x_1/x_2 + r_{12}x_3/r_{13}x_2}$$

Note that the terpolymerization can still be described in terms of
binary reactivity ratios.

Marconi et al. [2] have simulated binary copolymerization
where the required input data are reactivity ratios, mole fraction
of monomer, and the number of monomer units to be polymerized into
a single polymer chain. This program stores the polymer chain as
it is produced so that not only the copolymer composition can be
calculated, but also the sequence distribution and the moments of

the distribution can be computed. Furthermore, visual display of
the chain is possible, since the exact sequence can be stored and
reproduced. A flow chart for Marconi's program is shown in Fig. 2.

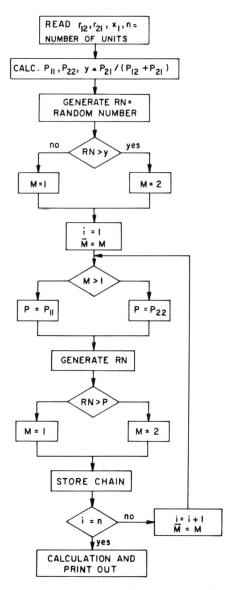

FIG. 2. Flow chart for simulation of binary copolymerization [2].

B. Sequence Distribution

Although his work dealt formally with homopolymer stereoregularity, Price [3] was the first to recognize the value of the Monte Carlo approach to visualizing the nature of a polymer chain. The mathematical description of a homopolymerization where the enchained monomer may be in either of two steric configurations is formally identical with a binary copolymerization. Price used as input data for his simulation the three fractions of triplets, iso-, syndio-, and heterotactic, which are obtainable from nmr analyses. These suffice to calculate the transition probabilities from a chain end of one configuration to that of another.

Price limited his chains to 101 units of length and as a consequence had considerable scatter in his results. Marconi et al. treated as many as 10^4 or more monomer units. The resulting precision of the Monte Carlo results is impressive when they are compared to expectation values for an infinite polymer chain. However, for most work, it appears that a chain of 500 to 1000 units is of adequate length to ensure statistical reliability.

Marconi et al. also dealt with the problem of a copolymerization where the monomer composition drifts with conversion. Such a system gives a copolymer composition distribution which has a breadth dependent on the magnitude of the drift. Analytical integration of the composition equation is cumbersome, and the Monte Carlo simulation can provide not only the composition distribution, but also the sequence distribution as a function of conversion.

Saito and Matsumura [4] have also used Monte Carlo simulation to determine sequence distributions. They investigated the problem for chlorinated polyethylene and for the analogous ethylene-vinyl chloride copolymer. Their input data consisted of triad probabilities determined from nmr data. However, such data give insufficient information on the chain when there are many chlorinated ethylene units separated by two or more unchlorinated units. Therefore, their input data also included a variable parameter which estimated the concentration of such runs of unchlorinated ethylene units; the calculation was iterated and the variable parameter modified until there was a convergence. This paper is of interest since the chlorinated polyethylene chain was simulated by a single "copolymerization" type of process, even though the experimental data came from both copolymerization and polymer modification processes.

C. Composition Distribution

Smidsrød and Whittington [5] have used a program similar to that of Marconi et al. [2] to describe the inhomogeneity in the composition of a copolymer sample which occurs because of short chain length. Their special concern is for the difficulty encountered in fractionation of biopolymers, such as polysaccharides, where the sample heterogeneity may be so severe that fractionation results may be misinterpreted as those of different entities. The problem is also of some importance in synthetic copolymer systems where incompatability may result if a sample has too broad a composition distribution. In the work of Smidsrød and Whittington, it is clear that interchain compositional heterogeneity decreases with increasing chain length but is still important for chains as long as 120 units. It is important to recognize the distinction between this type of heterogeneity, which occurs under constant conditions because the chains are too short to be at a statistical equiliibrium, and the heterogeneity which occurs in a high molecular weight copolymer produced over a range of conversion where the comonomer feed composition has drifted.

D. Reversible Copolymerization

O'Driscoll and co-workers in a series of papers [6] have dealt with the problem of copolymerization where the possibility of depropagation must be considered.

$$\sim\sim M_i^* + M_j \; \underset{\leftarrow}{\overset{\rightarrow}{}} \; \sim\sim M_i M_j^* \qquad (i, j = 1, 2)$$

The equations describing the composition of copolymers formed in the presence of depropagation could only be solved by a tedious numerical trial and error method. Therefore, a Monte Carlo simulation was employed. The polymer composition, sequence distribution and molecular weight were described by the simulation. Since the necessary kinetic parameters were not all experimentally accessible, this simulation proved valuable in that it was possible to use it to estimate the sensitivity of the copolymerization to various parameters. For example, in the cross-propagation reactions

$$\sim\sim M_i^* + M_j \rightleftarrows \sim\sim M_i M_j^* \quad (i, j = 1, 2; i \neq j)$$

the enthalpy change ΔH_{ij} is not experimentally accessible, but only the sum $(\Delta H_{ij} + \Delta H_{ji})$ is. By varying the relative magnitude of the two terms while keeping the sum constant, it was possible to see from Monte Carlo simulation [6b] that the temperature dependence of the composition behavior was strongly dependent on the relative magnitude. A comparison with experimental data [6c] suggested strongly that the assumption that $\Delta H_{ij} = \Delta H_{ji}$ was valid. The use of simulation for examining parametric sensitivity, as in this case, is an example of a unique value of simulation. It would have been impossible to survey the parameters experimentally, and, as mentioned, the analytical composition equations could only be solved numerically and with great difficulty.

A complete program for simulation of a terpolymerization with depropagation is presented as an appendix to this chapter.

III. CHAIN BRANCHING

One of the more complicated kinetic models which has been investigated by simulation is the free radical polymerization of ethylene where Wu et al. gave consideration to the "back-biting" mechanism for chain branching and termination [7]. In this mechanism short side chains are formed by abstraction of a hydrogen atom by the chain end from a carbon atom four atoms away from it, followed by monomer addition to the secondary or tertiary radical so formed. Chain termination occurs by scission of the carbon-carbon bond β to

a tertiary free radical. The multiplicity of structures which can
be formed is partially revealed by Fig. 3. A total of 53 such
structures were found to be necessary to represent nearly all of
the chain which was generated by reasonable values of the various
reaction probabilities. The Monte Carlo simulation used by Wu
et al. [7] provided for chain branching the same advantage as that
noted by Price [3] for stereoregularity: the opportunity to visu-
alize a polymer chain in all its structural chaos or order. While

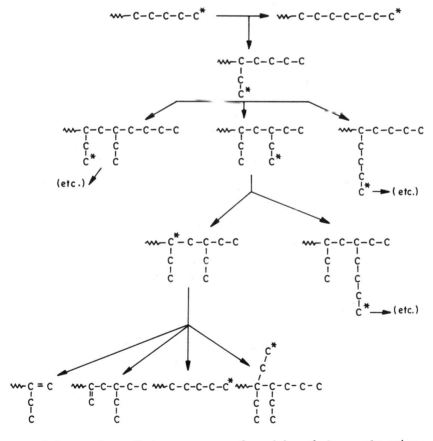

FIG. 3. Some of the structures formed in ethylene polymerization [3]. Each arrow represents the addition of a C = C unit.

Wu et al. regarded their simulation as somewhat unsatisfactory because of insufficient data to use in estimating some probabilities,
they did note that the simulation would be useful as a tool for
interaction with further experimental work.

IV. STEP-GROWTH POLYMERIZATION

All of the foregoing examples have dealt with vinyl, addition
polymerizations. The use of Monte Carlo simulation has also been
shown to be of value in polymerizations which follow a step growth
mechanism. Glasser and Glasser [8] have made a start on the simulation of the naturally occurring polymerization of lignin. They
assume that four mesomeric free radicals can be formed by the re-
moval of a hydrogen atom from coniferyl alcohol (Fig. 4). Polymerization is simulated by allowing the coupling of pairs of these
radicals and those derived by coupling reactions. The relative
probabilities of the many possible reactions have been assigned by
a judicious interpretation of existing knowledge on different types
of coupling plus the frequency of occurrence of known structures in
lignin and its derivatives. Obviously, such methods for assignment
of reaction probabilities in a simulation can only result in an incomplete model. The authors explicitly recognize this and point
out several improvements in the model which are needed.

Another step-growth polymerization which has been simulated is
that of the self-condensation of a monomer RA_f, where the functional

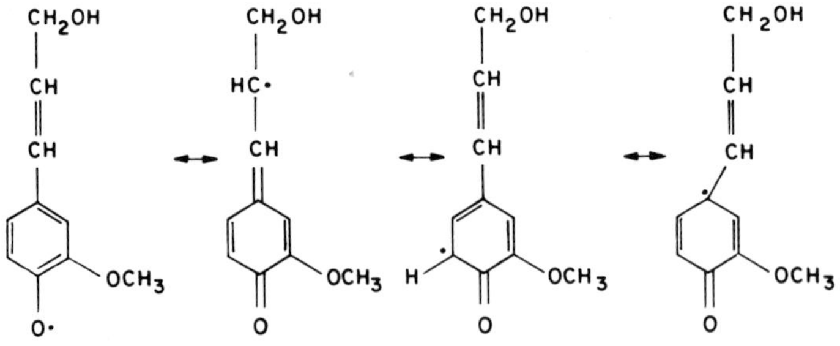

FIG. 4. Mesomeric radicals of coniferyl alcohol [8].

groups A combine to give a highly branched, possibly cyclized polymer (e.g., see Fig. 5). Interest in the molecular weight distributions which arise from such polymerizations, originally treated
theoretically by Flory [9], arose when Masson et al. [10] published
a different theoretical treatment and some data on silicic acid
polymers which sustained their treatment, albeit not unequivocally.
Given one theoretical model, two different theoretical treatments,
and equivocal experimental data, simulation is a perfect means of
resolving the difficulties. Falk and Thomas [11] have shown that
the simulation can not only resolve the differences between Flory
and Masson, but that it also shows the way to some unexpected results with respect to the occurrence of ring formation. In this
particular case, the rigor imposed by the logic necessary for a
successful simulation led to a significant gain in the understanding of polymerization where cyclization is possible.

V. DEGRADATION

Mention has been made of simulation of polymerization [1] and
copolymerization [6] where depropagation was considered. It is
also important to note several instances where Monte Carlo simulation of degradation has been achieved. The first such work was
that of Kotliar and Podgor [12], who followed a procedure proposed

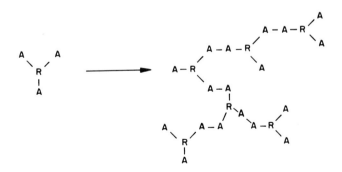

FIG. 5. One possible structure of hexamer formed by self-
condensation of RA_3.

by Kahn [13] for the decay of neutrons passing through a solid
medium. In the case of polymer degradation a given molecular weight
distribution is assumed or known from experimental data. Generation
of one random number selects a given length chain; if scission is to
occur, a second random number selects the position in the chain and
thus the size of the fragments formed, if cross-linking is to occur,
a second random number is used to determine the size of·the second
molecule and thus the size of the cross-linked product. Malac [14]
has applied this technique extensively to polyvinylchloride.

Meddings and Potter [15] have simulated the degradation of
amylopectin, a branched polyglucoside. Their simulation results
compared well with the results calculated from the analytical
solutions possible for the degradation of low molecular weight
analogues.

VI. SUMMARY

Monte Carlo simulations have been used on a variety of homo-
and copolymerizations, both for addition and for condensation polymerization. Linear and branched polymers have been treated; depropagation, chain scission and cross-linking have all (separately)
been simulated. This brief catalogue reveals that this technique
is useful in simulating any polymerization where the conditional
probability of each possible reaction for an active protion of the
polymer chain or chain end can be estimated.

These simulations are of particular value for envisioning the
structure of the chain as illustrated by Price [3] and by Wu et al.
[7]. They can be used to discriminate between different models as
shown by O'Driscoll and his co-workers [6] or to test theories as
shown by Falk and Thomas [10]. In an extension of their work on
binary, reversible copolymerizations to three component systems,
Kang and O'Driscoll have even used simulation for the design of
experiments [6d].

The cost of simulation must be reckoned in the light of the
cost of computer facilities available to the simulator. On a large

machine, a vinyl polymer chain 500 units long can be "synthesized" for about $1.00; however condensation polymerizations are perhaps an order of magnitude more expensive because of the indexing necessary. Although these costs may seem somewhat high in relation to the information content of a single polymer chain, there is often no alternate way of obtaining such information. In such a case, the value of simulation is inestimable, even though its cost is.

APPENDIX: Computer Program for Monte Carlo Simulation of Reversible Three-Component Polymerization

This program, taken from the Ph.D. Thesis of B. K. Kang, University of Waterloo, 1973 was used in Ref. [6d] of this chapter. It is presented here as an explicit example of the programming necessary to simulate polymerization. Using thermodynamic and kinetic data, terpolymerization is simulated at a succession of temperatures.

```
C     ***************************************************************
C     MONTE-CARLO SIMULATION OF REVERSIBLE THREE COMPONENT
C     POLYMERIZATION
C     ***************************************************************

      DIMENSION R(3,3),H(3),S(3),FII(3),EA(3),DK(3),NT(3),A(3),TX(10)
      DIMENSION P(3,3,4),XM(15,3),FS(3,3),RS(3,3),EK(3,3),LINE(1100)
      DIMENSION FI(3)
      REAL M(3)

C     ***************************************************************
C     DATA INPUT
C     ***************************************************************

      READ,H,S
      READ,FI,EA
      READ,((R(I,J),J=1,3),I=1,3)
      READ,LT
      READ,(TX(I),I=1,LT)
      READ,MNO
      READ,((XM(I,J),J=1,3),I=1,MNO)

C     ***************************************************************
C     PRINT OUT PARAMETERS USED
C     ***************************************************************

      PRINT 400,H,S
      PRINT 401,(FI(I),I=1,3),(EA(I),I=1,3)
      PRINT 500,((R(I,J),J=1,3),I=1,3)
      PRINT 102

C     ***************************************************************
C     TEMPERATURE DATA STORED IN TX ARE TRANSFERED TO T ONE PER LOOP
```

```
C
C
      *****************************************************************************************
      DO 1111  I1=1,LT
      T = TX(I1)
      T = T + 273.
      DO 32  I=1,3
   32 FII(I) = FI(I)*EXP(EA(I)/1.987*(1./333.1./T))
      *****************************************************************************************
C  EVALUATION OF CONSTANTS
      *****************************************************************************************
C
C  EVALUATION OF EQUILIBRIUM CONSTANT EK(I,J)
   10 DO 10  I=1,3
      DK(I) = EXP((T*S(I)-H(I))/(1.987*T))
      DO 11  I=1,3
      DO 11  J=1,3
   11 EK(I,J) = SQRT(DK(L)*DK(J))
C
C  EVALUATION OF FORWARD RATE CONSTANT FS(I,J)
      DO 13  I=1,3
      DO 13  J=1,3
   13 FS(I,J) = FII(I)/R(I,J)
C
C  EVALUATION OF REVERSE RATE CONSTANT RS(I,J)
      DO 14  I=1,3
      DO 14  J=1,3
   14 RS(I,J) = FII(I)/(R(I,J)*EK(I,J))
C
C  *****************************************************************************************
C  MONOMER CONCENTRATION STORED IN XM ARE TRANSFERED TO M
C  ONE SET PER LOOP
C  *****************************************************************************************
```

```
C
            DO 1111  J1=1,MNO
      33    DO 33  K1=1,3
            M(K1) = XM(J1,K1)

C     ***********************************************************************
C     EVALUATION OF PROBABILITIES P(I,J,K)
C     ***********************************************************************
C
            DO 15  I=1,3
            DO 15  J=1,3
            DO 15  K=1,3
            SUM = 0.0
            DO 16  KK=1,3
      16    SUM = SUM + FS(J,KK)*M(KK)
      15    P(I,J,K) = FS(J,K)*M(K)/(SUM + RS(I,J))
            DO 17  I=1,3
            DO 17  J=1,3
            SUM = 0.0
            DO 18  K=1,3
      18    SUM = SUM + FS(J,K)*M(K)
      17    P(I,J,4) = RS(I,J)/(SUM + RS(I,J))

C     ***********************************  ***********************************
C     INITIALIZATION
C     ***********************************************************************
C
            DO 40  I=1,1000
      40    LINE(I) = 0
            DO 19  I=1,5
      19    LINE(I) = 1
            I = 1
            J = 1
            N = 5
```

```
C *****************************************************************************
C      INVOKE MONTE-CARLO METHOD
C *****************************************************************************
       NCOUNT = 1
       NSTART = 9
       NR = 1
C
C      BUILT IN SUBPROGRAM RANDS GENERATE RANDOM NUMBERS
C      BETWEEN 0. AND 1.
   205 CALL RANDS(NSTART,X,NR)
       NCOUNT = NCOUNT + 1
C *****************************************************************************
C      DECIDE WHETHER TO PROPAGATE OR DEPROPAGATE
C *****************************************************************************
       K = 1
   202 SUM = 0.0
       SUM = SUM + P(I,J,K)
       IF(X.LE.SUM)GO TO(200,200,200,201),K
       K = K + 1
       GO TO 202
   200 N = N + 1
       LINE(N) = K
       I = LINE(N-1)
       J = LINE(N)
       GO TO 203
   201 N = N - 1
       IF(N.EQ.4)GO TO 51
       LINE(N+1) = 0
       I = LINE(N-1)
       J = LINE(N)
```

```
      GO TO 203
C
C     ****************************************************************
C     CONTROL ITERATION BY CONFINING MAXIMUM CHAIN LENGTH   N=1005
C     OR MAXIMUM NO. OF RANDOM EXPERIMENTS  NCOUNT=8000
C     ****************************************************************
   51 N = N + 1
  203 IF(N.GE.1005.OR.NCOUNT.GT.8000)GO TO 204
      GO TO 205
C
C     ****************************************************************
C     EVALUATION OF OUTPUTS
C     ****************************************************************
C
C     EVALUATION OF POLYMER COMPOSITION
  204 DO 21  I=1,3
   21 NT(I) = 0
      L = N - 5
      DO 20 I=1,L
      N = I + 5
      IF(LINE(N).NE.1)GO TO 22
      NT(1) = NT(1) + 1
   22 IF(LINE(N).NE.2)GO TO 20
      NT(2) = NT(2) + 1
   20 CONTINUE
      NT(3) = L - NT(1) - NT(2)
      DO 25  I=1,3
   25 A(I) = 100.0*FLOAT(NT(I))/FLOAT(L)
      SUM = 0.0
      DO 28  I=1,3
   28 SUM = SUM + M(I)
```

```
      DO 29  I=1,3
   29 M(I)/AUM*100.0
 1111 CONTINUE
  400 FORMAT(///,'1','DATA USED'//' ',3X,'ENTHALPY',27X,'ENTROPY'/' ',
     *3F10.1,5X,3F7.1)
  401 FORMAT(' ','FORWARD RATE CONST(60C)',15X,'ACTIVATION ENERGY'//' ',
     *3F10.1,5X,3F10.1)
  500 FORMAT(' ',3X,'REACTIVITY RATIO'/(' ',3X,3F10.2))
  102 FORMAT('0',' \EMP(C)',18X,'MONOMER MOLE-FRACTION',19X,'POLYMER
     * MOLE PERCENT ',10X,'N',7X,'NCOUNT'//)
      PRINT 103, T-273.,(M(I),I=1,3),(A(I),I=1,3),N,NCOUNT
  103 FORMAT(' ',F10.3,10X,3F10.3,10X,3F10.3,5X,14,5X,14/)
      STOP
      END
```

REFERENCES

1. G. G. Lowry, "Markov Chains and Monte Carlo Calculations in Polymer Science," Chap. 8, Marcel Dekker, Inc., New York, 1970.

2. P. F. Marconi, R. Tartarelli, and M. Capovani, *Chemica e l' industria*, Supplement, *7*, 1 (1971).

3. F. P. Price, *J. Polymer Sci.*, *C 25*, 3 (1968).

4. T. Saito and Y. Matsumura, *Polym. J.*, *4*, 124 (1973).

5. O. Smidstrdd and S. G. Whittington, *Macromol.*, *2*, 42 (1969).

6. (a) J. A. Howell, M. Izu, and K. F. O'Driscoll, *J. Polym. Sci.*, *A-1,8*, 699 (1970); (b) M. Izu and K. F. O'Driscoll, *J. Polym. Sci.*, *A-1,8*, 1675 (1970); (c) M. Izu and K. F. O'Driscoll, *J. Polym. Sci.*, *A-1,8*, 1687 (1970); (d) B. K. Kang and K. F. O'Driscoll, *Macromol.*, *7*, 000 (1974).

7. P. C. Wu, J. A. Howell, and P. Ehrlich, *Ind. Eng. Chem. Prod. Res. Develop.*, *11*, 35] (1972).

8. W. G. Glasser and H. R. Glasser, *Macromol.*, *7*, 17 (1974).

9. P. J. Flory, *Principles of Polymer Chemistry*, Chap. 9, Cornell University Press, Ithaca, New York, 1953.

10. (a) S. G. Whiteway, I. B. Smith, and C. R. Masson, *Can. J. Chem.*, *48*, 33 (1970); (b) C. R. Masson, I. B. Smith, and S. G. Whiteway, *Can. J. Chem.*, *48*, 201, 1456 (1970); *51*, 1422 (1973).

11. M. Falk and R. E. Thomas, *Can. J. Chem.*, *52*, 000 (1974).

12. (a) A. M. Kotliar and S. Podgor, *J. Polym. Sci.*, *55*, 423 (1961); (b) A. M. Kotliar, *J. Polym. Sci.*, *A1*, 3175 (1963).

13. H. Kahn, *Nucleonics*, *6(5)*, 27 (1950).

14. J. Malac, *J. Polym. Sci.*, *A-1,9*, 3563 (1971); *J. Polym. Sci.*, *C33*, 227 (1971); *J. Macromol. Sci. Chem.*, *7*, 923 (1973).

15. P. J. Meddings and O. E. Potter, *Adv. Chem. Series*, *109*, 96 (1972).

Chapter 4

CALCULATION OF MONOMER REACTIVITY RATIOS FROM MULTICOMPONENT COPOLYMERIZATION RESULTS

Alfred Rudin
University of Waterloo
Departments of Chemistry and Chemical Engineering
Waterloo, Ontario, Canada

I. INTRODUCTION

This chapter reviews methods for computing copolymerization reactivity ratios in general and the use of multicomponent copolymerization data in particular. It is assumed that the reader is familiar with the basic theory and important developments in copolymerization. Suitable background material is presented in several well-known special sources [1-3], as well as in most textbooks on polymer chemistry.

Our concern here is mainly with a computer-assisted method for
calculating reactivity ratios directly from results of multicomponent copolymerization experiments. Such systems are generally of
more practical interest than binary copolymerizations, since many
important commercial copolymers are made by combining three or more
monomers.

A number of copolymerization models purport to link the monomer feed composition and corresponding copolymer composition. This
review is confined to use of the so-called "simple copolymer theory"
[4-6] because this representation has fewer parameters than alternative models and the following presentation is therefore least
cumbersome. All the multicomponent copolymerizations which the
author has investigated can be analyzed satisfactorily with the
simple copolymer model. Extension of the computer analyses described below to more complicated copolymerization models appears to be
straightforward.

The examples cited in this article pertain to free radical
copolymerizations. The concepts employed are usually applicable,
however, to copolymerizations with other modes of initiation.

All multicomponent copolymerization theories employ binary reactivity ratios. The composition of the copolymer formed from n
monomers can be expressed, for example, in terms of the corresponding instantaneous monomer feed composition and n(n - 1) binary reactivity ratios, according to the simplest copolymer theory [6].
These reactivity raios have been measurable heretofore only in binary copolymerizations, and it has been necessary to assume that
such binary ratios apply unchanged in multicomponent reactions.

The weight of evidence indicates that this assumption is probably true in free radical copolymerizations [6-9]. That is to say,
measured copolymer compositions have been thought to agree with
values predicted from binary reactivity ratio calculations to within the rather large uncertainties which are unfortunately associated
with copolymerization experiments [10]. There are, however, several
exceptions to this conclusion [11,12], which suggest the desirability

of measuring reactivity ratios directly in multicomponent copolymerizations where this is feasible.

Two other considerations reinforce the value of reactivity
ratio calculations based on measurements in particular working systems. The experimental labor involved is considerably reduced if a
series of multicomponent compolymerizations is studied instead of
n(n - 1) binary systems. Also, recent studies [13-15] have shown
that propagation rates and reactivity ratios in free radical polymerizations may be influenced by the nature of the solvent (or
monomer) medium. If these effects are found to be generally significant, it will be all the more prudent to use reactivity ratios
which apply to particular working systems. Such reactivity ratios
are most efficiently and easily determined by direct study of multicomponent copolymerizations under practical reaction conditions.

It has become apparent in recent years that the methods used to
analyze experimental results may have a strong influence on the magnitudes and reliability of reactivity ratios calculated from binary
copolymerizations. This general conclusion applies with equal force
to estimates of reactivity ratios in multicomponent systems. This
chapter therefore begins with a brief review of the pertinent literature pertaining to binary system calculations. Computer programs
are referenced but are not listed in detail in this connection because they are not directly connected with the main thrust of this
article.

II. GENERAL COMMENTS ON COMPUTER-ASSISTED CALCULATION
PROCEDURES FOR REACTIVITY RATIOS

Many of the reactivity ratios available in the literature were
calculated from linearized versions of the simple copolymer equation
[4]:

$$\frac{dM_1}{dM_2} = \frac{M_1(r_1M_1 + M_2)}{M_2(M_1 + r_2M_2)} \tag{1}$$

where dM_1 and dM_2 are the decrements in concentrations of monomers
1 and 2 which are present in the feed in concentrations M_1 and M_2,

respectively. The reactivity ratios r_1 and r_2 in Eq. (1) are defined, as usual, as the ratios of rate constants for addition of a
particular radical end to the same or dissimilar monomer [1].

The most widely used linearization method is that of Fineman
and Ross [16], which produces a convenient graphical solution for
r_1 and r_2 in terms of the slope and intercept in a rectilinear plot.
Linear least-squares fitting of the data points is employed, and
standard deviations may be calculated for the slopes and intercepts,
and hence for the reactivity ratios. These estimates of reliability
may be misleading, however, because the observed copolymer composition is contained in both the "independent" and "dependent" variables in the linear equations [10]. The assumptions inherent in
linear least-squares regression analysis result also in heavy
weightings being assigned inadvertently to the experimental data
from extremes of the monomer feed compositions. The calculated reactivity ratios often differ, depending on the assignment of r
values between slope and intercept [8,10]. The Fineman-Ross method
is convenient and mathematically correct but its application to experimental copolymerization data is not statistically sound. The
reasons for this conclusion have been presented in detail by Tidwell
and Mortimer [17] and Behnken [18] has discussed the general errors
which may arise in linear least-squares treatment of copolymerization data.

The linearization method of Yezrielev and co-workers [19] employs a symmetrical transformation of the differential copolymer
equation. The values obtained for reactivity ratios do not differ
when the monomer order is reversed, as may happen in Fineman-Ross
calculations. Joshi [20] has itemized a computer program for application of this computational method and its extension to an integrated copolymer equation. This is stated [20] to be the best
linear computational method, although it seems not to be as reliable
in principle or in practice as the Tidwell-Mortimer [17] nonlinear
least-squares estimation which is mentioned below.

Equation (1) can be rearranged, as described by Mayo and Lewis

[4], to permit a graphical solution for r_1 and r_2 from the intersection region of straight line plots of the experimental data. The
method has been computerized and the subjective choice of the "best"
r_1 r_2 pair eliminated by employing a linear least-square solution
weighted according to the slopes of the various lines in the Mayo-
Lewis intersection plot [21]. This method has been extended to the
integral form of the copolymer equation [Eq. (1) is the differential
form], but the results are disappointing [20] presumably again because of the linearization involved in the method of intersections.

A useful computational method, which is not practical without
the use of a computer, involves the direct curve fitting of copoly-
mer-monomer compositions according to Eq. (1). This method is not
novel in principle [1], but its effective use requires more iterations than can be performed by manual calculation. Johnston and
Rudin [22] found that such an estimation was preferable to results
of calculations based on linearized forms of the differential copolymer equation. Braun and co-workers [23] have recently given
details of a FORTRAN IV program for curve-fitting computations of
binary reactivity ratios.

Although it is not so described by its authors, the nonlinear
least-squares computational method of Tidwell and Mortimer [17] is
also essentially a computerized curve-fitting calculation. The copolymer composition equation which is fitted is in the form due to
Skeist [24], and successive iterations for r_1 and r_2 are performed
with a variation of the Gauss-Newton nonlinear least-squares technique [25] to minimize the sum of squared deviations of experimental
and predicted data points. Some possible improvements in the computational method have been suggested recently by Joshi [20]. The
authors [17] claim that this is the best method developed to date
for estimation of reactivity ratios from binary copolymerization
data. Others who have tested the technique concur in this conclusion [20,22,26].

The three curve-fitting programs which have been mentioned
should result in essentially equivalent reactivity ratios from a

given set of experimental data. The Tidwell-Mortimer [17] procedure appears, however, to make more efficient use of computer time
than the techniques of Johnston and Rudin [22,26] or Braun and co-
workers [23].

Another important recent contribution is the provision of a
good measurement of the precision of estimated reactivity ratios.
The calculation of independent standard deviations for each reactivity ratio obtained by linear least-squares fitting to Fineman-
Ross or other linear forms of the differential copolymer equations
is invalid because the two reactivity ratios are not statistically
independent. Information about the precision of reactivity ratios
which are determined jointly is properly conveyed by specification
of joint confidence limits within which the true values can be
assumed to coexist. This is represented as a closed curve in a
plot of r_1 and r_2.

Standard statistical techniques for such computations are impossible or too cumbersome for application to binary copolymerization data in the usual absence of estimates of reliability of the
values of monomer feed and copolymer composition data. Tidwell and
Mortimer [17] have, however, provided a computer-assisted method
for the approximate estimate of such joint confidence loops.

This method has not been extended to the assessment of the
precision of reactivity ratios calculated directly from multicomponent copolymerization data. It is nevertheless of value in the
present context because it can be used to compare "binary" and
"multicomponent" reactivity ratios. Rudin and co-workers [27] have
used such comparisons to justify exclusion of certain raw data in
the computation of reactivity ratios in three component systems.
When binary reactivity ratios are available joint confidence limit
calculations can also be employed to determine whether the nature
of the multicomponent polymerization medium has caused significant
deviations from behavior in simpler systems.

Attention has been focused in this review on computation of
reactivity ratios by the fitting of corresponding feed and copolymer

compositions to differential forms of the copolymer equation. It is clear that the most reliable results are obtained by nonlinear least-squares optimizations of these data. Terpolymerization results are considered in detail below. A differential form of the simple copolymer model is used in the multicomponent case and non-linear optimization is applied to the experimental values to calculate reactivity ratios.

Mention should also be made of the potential advantages of using an integrated form of the copolymer equation to estimate reactivity ratios from binary copolymerization data. These benefits are particularly significant when the relative monomer concentrations may change appreciably during the course of the experiment [28]. Montgomery and Fry [29] and Harwood and co-workers [30] have published computer programs for calculation of reactivity ratios from the integrated copolymer equation in binary systems.

It should be pointed out in this connection the use of the integrated copolymerization equation is necessary only when experiments have been carried to a relatively high conversion. Reactivity ratios calculated by inserting the arithmetic means of initial and final monomer feeds into the differential equation coincide closely with those from integral curves when conversions are limited to about 20% or less [20]. The mathematical computations are simpler, of course, when the differential equation can be used.

The copolymer composition can be estimated usefully in many cases from the composition of unreacted monomers, as measured by gas-liquid chromatography (GLC). Analytical errors are reduced if the reaction is carried to as high a conversion as possible, since the content of a given monomer in the copolymer equals the difference between its initial and final measured contents in the feed mixture. The uncertainty in the copolymer analysis is thus a smaller proportion of the estimated quantity, the greater the magnitude of the decrease in the monomer concentration in the feed [31]. It may seem appropriate under these circumstances to estimate reactivity ratios by fitting the data to an integrated form of the copolymer equation.

The raw GLC data are usefully smoothed by plotting the unreacted weight of each monomer in a given initial mass of monomers against reaction time. Such plots are usually linear within experimental uncertainty up to about 10 to 20% conversion. The extent of reaction at which curvature becomes pronounced depends, of course, on the relative magnitudes of the reactivity ratios, except for azeotropic feed mixtures. A linear experimental relation shows that the copolymer composition does not vary significantly. Both the differential and integrated forms of the copolymer equation are then equally reliable in this range of monomer conversions [9].

Copolymer composition in multicomponent systems can be estimated with given reactivity ratios by computer-assisted calculations based on integrated and differential feed-polymer composition equations [8,9]. The reverse operation, in which reactivity ratios are calculated from multicomponent copolymerization data and which is discussed in detail below, is practical at present only with differential forms of the copolymer equation. This requires the use of low conversion data (up to about 10% conversion, depending on the particular case) or averaged initial and final feed composition values. It is unlikely that measurements taken to high conversions will be amenable to the present analysis.

III. ESTIMATION OF REACTIVITY RATIOS FROM MULTI-COMPONENT COPOLYMERIZATION DATA

As mentioned above, reactivity ratios are usually determined experimentally by fitting corresponding monomer feed and polymer compositions in binary copolymerizations to an appropriate kinetic model, such as the simple differential copolymer formula given in Eq. (1). Estimation of reactivity ratios directly from feed and copolymer compositions in multicomponent reactions is not practical without the use of a computer. The adequacy of binary reactivity-ratio values and of the copolymerization model has been judged subjectively from the agreement between experimental and calculated copolymer compositions. The advantages of direct use of multicomponent polymerization data for estimation of reactivity ratios have been enumerated earlier in this article.

The number of feed compositions in an experimental study of copolymerization is usually quite small, because of the time-consuming nature of such studies. The uncertainty in copolymer compostion analyses may be appreciable [9,10]. It seems to be impractical, therefore, to derive reliable reactivity ratios from analytical solutions of the multicomponent copolymer equation in parallel with some of the methods cited above which may be applied to binary systems. The technique described here estimates multi-component reactivity ratios by minimizing deviations between predicted and observed copolymer compositions. We have already noted that this is also the preferred approach for binary copolymerization studies.

The program given below produces best-fit reactivity ratios, given copolymer and corresponding monomer feed compositions and preliminary values of the reactivity ratios. The accuracy of such preliminary values is not critical. Q o calculations [32,33] could be used, for example, if experimental binary system values are lacking.

Three computer methods have been described for optimizing reactivity ratios from ternary copolymerization results [27]. The most efficient of these procedures is given here. Extension to reactions with four or more monomers is straightforward.

The instantaneous copolymer composition is given in terms of decrements in monomer feed concentration, dM_i, by the differential equation of Alfrey and Goldfinger [1]:

$$\frac{dM_1}{dM_3} = \frac{M_1(M_1r_{23}r_{32} + M_2r_{31}r_{23} + M_3r_{32}r_{21})(M_1r_{12}r_{13} + M_2r_{13} + M_3r_{12})}{M_3(M_1r_{12}r_{23} + M_2r_{13}r_{21} + M_3r_{12}r_{21})(M_3r_{31}r_{32} + M_1r_{32} + M_2r_{31})}$$

(2)

$$\frac{dM_2}{dM_3} = \frac{M_2(M_1r_{32}r_{13} + M_2r_{13}r_{31} + M_3r_{12}r_{31})(M_2r_{21}r_{23} + M_1r_{23} + M_3r_{21})}{M_3(M_1r_{12}r_{23} + M_2r_{13}r_{21} + M_3r_{12}r_{21})(M_3r_{31}r_{32} + M_1r_{32} + M_2r_{31})}$$

(3)

where the M_i are the molar concentrations of monomer i (i = 1, 2, 3) and r_{ij} (i $\neq$ j) are the appropriate binary reactivity ratios.

In the computations of r_{ij} a linear search can be conducted
along any direction with an accelerating stepsize and a quadratic-
smoothing feature to obtain more accuracy. The search proceeds
along the chosen direction with stepsize at the ith trial being
2^{i-1} h (for suitably chosen h) until the minimum in that direction
has been bracketed. The bracketing interval is then searched with
decreasing stepsize until the minimum is known to lie in an interval of length 2h. A quadratic fit is then performed on the three
points closest to the minimum, and the minimum of this quadratic
is chosen as the new minimum. If the quadratic fit has resulted
in a significant improvement, it is repeated once more using the
three best points.

Each cycle of this optimization involves first a search along
the first coordinate axis. If a point is found which is significantly different from the starting point, a search along the second
axis is begun from this new point. If no such point is found, the
search along the second axis starts at the original point. A successive search of each axis in turn is thus carried out either
until all six have been found to give no improvement or until a
significant improvement has been found along two of them. In the
former case the search is terminated, and in the latter a new
search is conducted along the direction defined by a line through
the original starting point and the new minimum point. The point
resulting from this search is then treated as a new starting value,
and beginning there, the remaining coordinate axes are searched as
above.

The new direction to be searched is defined by

$$H = (H_1 \cdots H_n) \frac{x_i^* - x_i^o}{\left[\sum_{i=1}^{n} (x_i^* - x_i^o)^2 \right]^{1/2}}$$

where x_i^o is the ith component of the original point and x_i^* is the
i^{th} component of the minimum point obtained. Because a new direction is searched every time a significant move is detected along

any two axes, only two of the x_i^o will differ from the corresponding x_i^*.

A. Flow Chart

Figure 1 shows the flow chart for one cycle of this procedure. The program is printed out following the description of the flow chart.

The original reference [27] contains examples of the application of this computational method to experimental results in the free radical copolymerizations of systems containing styrene/α-methylstyrene/methacrylonitrile, butadiene/styrene/2-methyl-5-vinylpyridine and acrylonitrile/methyl methacrylate/α-methylstyrene.

Description of Flow Chart (see Fig. 1)

(1) - (2)   The data, initial estimates, and search parameters are read in.

(3)   The vector ξ, giving the direction of the search, is set to zero, then i is set to one to begin the search along the first axis.

(4)   The linear search is conducted along ξ from $\underline{x}^o$ to yield $\bar{x}$.

(5)   A check is made to determine if the minimum point has moved significantly. If it has not, one proceeds to set up the search along the next coordinate axis (6) .

(7)   This check is made only if the minimum did move significantly. Its purpose is to determine if this is the first or second time such a move has occurred. If it is the first, one proceeds to (8) , otherwise to (9) .

(8)   The original minimum x^o is stored in XK, and x^o is then updated to contain the minimum from the first search. The dimension counter ID is incremented and one proceeds to set up the search along the next axis (6) .

(9) - (11)   The new direction is defined (9) and the starting point is updated (10) before the search is conducted (11) .

(12) - (14)   The new minimum is checked; if it has moved significantly, we store the new value (13) . If not, we retain the old one (14) .

(15) - (16)   The direction vector is reset to search the next coordinate axis.

(7)   If all axes have been searched once, the cycle terminates and prints results (17) .

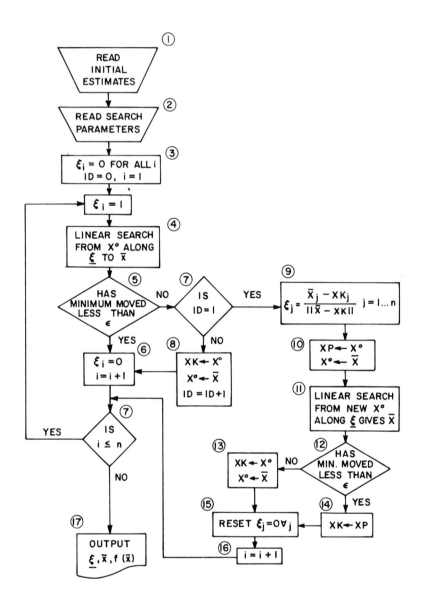

FIG. 1.   Flow Chart.

B. Computer Program

```
$JOB      WATFIV  **********,I=(0,10),P=25
 1        IMPLICIT REAL*8 (A-H,O-Z)
 2        COMMON XM(3,15),RA(2,15),RER(2,15),NFV,NEXPT
 3        DIMENSION RR(3,15)
 4        DIMENSION ETA(10),XO(10),XBAR(10),H(6),Y(10),X1(10),X(10)
 5        DATA H/6*0.01D0/,NFV/0/
    C ***** READ IN VALUES'
 6        READ,NEXPT,ITER
 7        DO 4 I=1,NEXPT
 8      4 READ,(XM(J,I),J=1,3)
 9        DO 5 I=1,NEXPT
10      5 READ,(RR(J,I),J=1,3)
11        READ,(XO(I),I=1,6)
12        DO 6 I=1,NEXPT
13        RER(1,I)=RR(1,I)/RR(3,I)
14      6 RER(2,I)=RR(2,I)/RR(3,I)
    C ***** N IS DIMENSION OF XO
15        N=6
16        EPS=.1D-20
17        PRINT,H
18        DO 77 IJK=1,ITER
19        CALL NCYC( XO  ,EPS,N,H,XBAR)
20        PRINT,' FUNCTION EVALUATIONS =', NFV
21        DO 77 IKL=1,6
22     77 XO(IKL)=XBAR(IKL)
23        STOP
24        END

25        DOUBLE PRECISION FUNCTION F(R)
26        IMPLICIT REAL*8(A-H,O-Z)
27        COMMON XM(3,15),RA(2,15),RER(2,15),NFV,NEXPT
```

```
      DIMENSION XN1(15),XN2(15),XD1(15),R(6)
      F=0.0D0;NFV=NFV+1
C.....EVALUATE EXPECTED RATIOS
      DO 1 I=1,NEXPT
      XN1(I)=XM(1,I)*(XM(1,I)*R(5)*R(6)+XM(2,I)*R(4)*R(5)+XM(3,I)*R(6)*
     1R(2))*(XM(1,I)*R(1)*R(3)+XM(2,I)*R(3)+XM(3,I)*R(1))
      XD1(I)=XM(3,I)*(XM(1,I)*R(5)+XM(2,I)*R(1)*R(5)+XM(2,I)*R(3)*R(2)+XM(3,I)*R(1)*
     UR(2))*(XM(3,I)*R(4)*R(6)+XM(1,I)*R(6)+XM(2,I)*R(4))
     1 XN2(I)=XM(2,I)*(XM(1,I)*R(6)*R(3)+XM(2,I)*R(3)*R(4)+XM(3,I)*R(1)*
     IR(4))*(XM(2,I)*R(2)*R(5)+XM(1,I)*R(5)+XM(3,I)*R(2))
      DO 2 I=1,NEXPT
      RA(1,I)=XN1(I)/XD1(I)
    2 RA(2,I)=XN2(I)/XD1(I)
C.....CALCULATE LEAST SQUARE DIFFERENCE
      DO 3 I=1,NEXPT
    3       F= F+(RER(1,I)-RA(1,I))**2+(RER(2,I)-RA(2,I))**2
      RETURN
      END
      SUBROUTINE NCYC (XO,EPS,N,H,XBAR)
      IMPLICIT REAL*8(A-H,O-Z)
      DIMENSION ETA(6),Y(6),XBAR(6),X1(6)  ,XO(N),XP(6),XK(6),H(6)
C ***** INITIALIZATION
      PRINT101
      ID=0
      DO 1 I=1,N
    1 ETA(I)=0.D 00
C ***** START SEARCH
      I=1
    2 ETA(I)=1.D 00
      CALL BERMAN(ETA,XO,Y,H,N,XBAR,X1)
      XM=0.D 00
      DO 3 J=1,N
```

Line numbers: 28, 29, 32, 33, 34, 35, 36, 37, 38, 39, 40, 41, 42, 43, 44, 45, 46, 47, 48, 49, 50, 51, 52, 53

```
54      3 XM=XM+(Y(J)-XBAR(J))**2
55        XM=DSQRT(XM)
56      C ***** CHECK IF WE HAVE MOVED SIGNIFICANTLY
57        IF(XM.LT.EPS)GO TO 44
58        IF(ID.EQ.1)GO TO 5
59        DO 6 J=1,N
60        XK(J)=XO(J)
61      6 XO(J)=XBAR(J)
62        ID=1
63        GO TO 4
64     44 PRINT102,EPS,(ETA(J),J=1,N)
      102 FORMAT(1HO,'MINIMUM MOVED LESS THAN ',D14.6/' ALONG ETA =',6D14
          #)
65      4 ETA(I)=0.D 00
66        I=I+1
67     15 IF(I.LE.N)GO TO 2
68        PRINT100,(XBAR(J),J=1,N)
69        GO TO 16
70      5 XM=0.D 00
71        DO 7 J=1,N
72      7 XM=XM+(XK(J)-XBAR(J))**2
73        XM=DSQRT(XM)
74        DO 8 J=1,N
75        ETA(J)=(XBAR(J)-XK(J))/XM
76        XP(J)=XO(J)
77      8 XO(J)=XBAR(J)
78        CALL BERMAN (ETA,XO,Y,H,N,XEAR,X1)
79        XM=0.D 00
80        DO 9 J=1,N
81      9 XM=XM+(XO(J)-XBAR(J))**2
82        XM=DSQRT(XM)
83        IF(XM.GT.EPS)GO TO 10
84        DO 11 J=1,N
```

```
85   11 XK(J)=XP(J)
86      GO TO 12
87   10 DO 13 J=1,N
88      XK(J)=XO(J)
89   13 XO(J)=XBAR(J)
90   12 DO 14 J=1,N
91   14 ETA(J)=0.D 00
92      I=I+1
93      GO TO 15
94   16 RETURN
95  100 FORMAT (1H ,'FINAL X VALUE FROM NON-CYCLIC SEARCH IS  '6D15.6)
**WARNING** EXPECTING COMMA BETWEEN FORMAT ITEMS NEAR S '6D
96  101 FORMAT(' ********** ENTERING NON-CYCLIC SEARCH  ********** ')
97      END

98      SUBROUTINE BERMAN (ETA,XO,Y,H,N,XBAR,X1)
99      IMPLICIT REAL* 8 (A-H,O-Z)
100     REAL*8 Y(N),XK(3,6),FK(3)
101     REAL*8 XO(N),X1(N),XBAR(N),ETA(N),H(6),DIREC(6)
102     INTEGER REV
    C
    C ***** INITIALIZATION
    C
    C ***** INITIALIZE QUADRATIC FIT COUNTERS
103     ID=3
104     DO 54 J=1,N
105     DO 54 JJ=1,3
106  54 XK(JJ,J)=1.D 06
107     FK(1)=0.D 00
108     FK(2)=0.D 00
109     FK(3)=0.D 00
110     DO 400 J=1,N
111 400 DIREC(J)=H(J)
```

```
      C
112         REV=0
113         BASE=2.0D 00
114         INDEX=0
115         F0=F(X0)
116         DO 50 J=1,N
117         Y(J)=X0(J)
118   50    XK(2,J)=X0(J)
119         FK(2)=F0
120         K=1
121   1     I=1
      C
      C ***** CALCULATE NEW X,F
      C
122         Do 3 J=1,N
123   3     X1(J)=X0(J)+DIREC(J)*ETA(J)*BASE**INDEX
124         F1=F(X1)
125         K=K+1
      C ***** CHECK IF NEW VALUE BETTER
126         IF(F1.LT.F0)GO TO 4
      C ***** WANT TO START ACCELERATION
127         IF(I.EQ.1)GO TO 5
      C ***** REVERS LOOP
128         IF(INDEX.EQ.0)GO TO 98
      C ***** STARTING VALUE IS BEST
      C
      C ***** ACCELERATION LOOP
      C
129   4     INDEX=INDEX+1
130         I=I+1
      C ***** CHANGE VALUES
131         DO 10 J=1,N
```

```
132     10 X0(J)=X1(J)
133        F0=F1
134     C ***** CALCULATE NEW VALUES
           DO 11 J=1,N
135     11 X1(J)=X0(J)+DIREC(J)*BASE**INDEX*ETA(J)
136        F1=F(X1)
137        K=K+1
        C ***** CHECK IF NEW VALUE IS BETTER
138        IF(F1.LT.F0)GO TO 4
        C ***** OTHERWISE START DECELERATION
        C
        C ***** DECELERATION LOOP
        C
139     21 IF(INDEX.EQ.0) GO TO 98
140        REV=0
141        INDEX=INDEX-1
142        IF(INDEX.NE.0)GO TO 27
        C ***** STORE VALUE FOR POSSIBLE QUADRATIC FIT
143        DO 70 KK=1,3
144        DO 70 J=1,N
145     70 XK(KK,J)=1.D 06
146        DO 52 J=1,N
147     52 XK(2,J)=X0(J)
148        FK(2)=F0
        C
149     27 KD=1
150        I=I+1
        C ***** CALCULATE NEW VALUES
151        DO 22 J=1,N
152     22 X1(J)=X0(J)+DIREC(J)*ETA(J)*BASE**INDEX
153        F1=F(X1)
154        K=K+1
155        IF(F1.LT.F0) GO TO 23
```

```
156   C ***** F1 BETTER-KEEP ON GOING IN SAME DIRECTION WITH SAME STEP SI
            IF(KD.EQ.1) GO TO 24
157   C ***** WORSE VALUE ON FIRST STEP - LOOK ON OTHER SIDE I.E. REVERSE
            GO TO 21
      C
      C ***** LINEAR SEARCH
      C
      C ***** CHANGE VALUES
158   23    IF(INDEX.NE.0)GO TO 123
      C ***** SHIFT VALUES FOR QUADRATIC FIT
159   123   CALL SHIFT (DIREC,ID,XK,FK,N)
160         DO 25 J=1,N
161         X0(J)=X1(J)
162   25    XK(ID,J)=X1(J)
163         FK(ID)=F1
      C
164         F0=F1
      C ***** CALCULATE NEW VALUES
165         I=I+1
166         KD=KD+1
167         DO 26 J=1,N
168   26    X1(J)=X0(J) + DIREC(J)*ETA(J)*BASE**INDEX
169         F1=F(X1)
170         K=K+1
171         IF(F1.LT.F0) GO TO 23
      C ***** REPEAT LINEAR SEARCH
172         IF(INDEX.NE.0)GO TO 21
173         CALL SHIFT (DIREC,ID,XK,FK,N)
174         DO 62 J=1,N
175   62    XK(ID,J)=X1(J)
176         FK(ID)=F1
177         GO TO 98
      C ***** DECREASE INDEX
```

```
      C
      C ***** REVERSE LOOP FOR DECELERATION
      C ***** STORE VALUE FOR POSSIBLE QUADRATIC FIT
178      24 IF(INDEX.NE.0)GO TO 124
179         ID=3
180         IF(DIREC(1).LT.0)ID=1
181         DO 53 J=1,N
182      53 XK(ID,J)=X1(J)
183         FK(ID)=F1
      C
184     124 REV=REV+1
185         IF(REV.NE.1) GO TO 21
186         DO 401 J=1,N
187     401 DIREC(J)=-DIREC(J)
188         GO TO 27
      C
      C ***** REVERSE LOOP
      C ***** STORE VALUE FOR POSSIBLE QUADRATIC FIT
189       5 ID=3
190         IF(DIREC(1).LT.0)ID=1
191         DO 51 J=1,N
192      51 XK(ID,J)=X1(J)
193         FK(ID)=F1
      C
194         REV=REV+1
195         IF(REV.NE.1) GO TO 98
196         DO 402 J=1,N
197     402 DIREC(J)=-DIREC(J)
198         GO TO 1
      C
      C ***** FINAL VALUE
```

```
199   C
200  98     DO 99 J=1,N
201         XBAR(J)=XO(J)
202  99     XO(J)=Y(J)
203         PRINT103
204         DO 59 J=1,3
205  59     PRINT104,FK(J),(XK(J,JJ),JJ=1,N)
206         EPS = .00001
207         CALL QUAD(XK,FK,N,EPS,H,ETA,XBAR)
208         RETURN
209  103    FORMAT(1H0,' BRACKETING VALUES'/' FUNCTION VALUE          X VALUE')
210  104    FORMAT(1H ,D14.7,6D14.5)
            END

211         SUBROUTINE SHIFT (DIREC,ID,XK,FK,N)
212         IMPLICIT REAL*8(A-H,O-Z)
213         DIMENSION FK(3),XK(3,6),DIREC(6)
214         IF(DIREC(1).LT.0)GO TO 55
215         ID=3
216         IF(XK(3,1).EQ.1.D 06)GO TO 123
217         DO 56 J=1,N
218         XK(1,J)=XK(2,J)
219  56     XK(2,J)=XK(3,J)
220         FK(1)=FK(2)
221         FK(2)=FK(3)
222         GO TO 123
223  55     ID=1
224         IF (XK(1,1).EQ.1.D 06)GO TO 123
225         DO 58 J=1,N
226         XK(3,J)=XK(2,J)
227  58     XK(2,J)=XK(1,J)
228         FK(3)=FK(2)
229         FK(2)=FK(1)
```

```
230      123 RETURN
231          END

232          SUBROUTINE QUAD(X,FK,N,EPS,HV,ETA,XR)
233          IMPLICIT REAL*8 (A-H,O-Z)
234          DIMENSIONX(3,6),FK(3),XQ(6),ETA(N),XQ2(6),XB(6),XR(N),HV(6)
235          H=0.D0
236          DO 78 J=1,N
237       78 H=H+HV(J)**2
238          H=DSQRT(H)
239          DO 77 J=1,N
240       77 XB(J)=X(2,J)
     C CHECK FOR LINEARITY
241          IF(((FK(1)-FK(2)).GT.EPS).OR.((FK (3)-FK(2)).GT.EPS))GO TO 1
     C IF LINEAR PRINT AND STOP
242          PRINT 100
243          PRINT106,FK(2),(X(2,J),J=1,N)
244          GO TO 197
     C CALCULATE HBAR FOR QUAD FIT
245        1 HBAR=H*(FK(1)-FK( 3))/(2.D 00*(FK(1)+FK(3)-2.D 00* FK(2)))
     C CALCULATE QUAD MIN
246          DO 2 J=1,N
247        2 XQ(J)=X(2,J)+HBAR*ETA(J)
     C FIND MODULUS OF DIFFERENCE
248          XDM=0.D 00
249          DO 3 J=1,N
250        3 XDM=XDM+(XQ(J)-X(2,J))**2
251          XDM=DSQRT(XDM)
     C CHECK SIZE OF MODULUS
252          IF(XDM.GT.EPS)GO TO 4
     C IF NO IMPROVEMENT,PRINT,STOP
253          PRINT101,EPS
254          PRINT106,FK(2),(X(2,J),J=1,N)
```

```
255         GO TO 197
256       4 DEL1=XDM
257         FXQ=F(XQ)
258         FXB=FK(2)
    C CHECK SIGN OF HBAR
259         IF(HBAR.LT.0.D 00) GO TO 5
    C CHECK FOR LINEARITY FOR SECOND FIT
260         IF((DABS(FK(3)-FXQ).GT.EPS).OR.(DABS(FK(2)-FXQ).GT.EPS)) GO TO 6
    C IF LINEAR PRINT AND STOP
261       7 PRINT102,DEL1
262         PRINT106,FXQ ,(XQ(J ),J=1,N)
263         GO TO 198
264       6 IF(FK(2).GT.FXQ)GO TO 8
    C SHIFT BRACKETS
265         FK(3)=FXQ
266         A=-H
267         C=HBAR
268         GO TO 10
269       8 FK(1)=FK(2)
270         FK(2)=FXQ
271         DO 12 J=1,N
272      12 X(2,J)=XQ(J)
273         A=-HBAR
274         C=H-HBAR
275         GO TO 10
    C CHECK FOR LINEARITY FOR SECOND FIT
276       5 IF((DABS(FK(1)-FXQ).LE.EPS).AND.(DABS(FK(2)-FXQ)).LE.EPS)GO TO 7
277         IF(FK(2).LT.FXQ)GO TO 11
    C SHIFT BRACKETS
278         FK(3)= FK(2)
279         FK(2)=FXQ
280         DO 14J=1,N
281      14 X(2,J)=XQ(J)
```

```
282        A=-H-HBAR
283        C=-HBAR
284        GO TO 10
285     11 FK(1)=FXQ
286        A=HBAR
287        C=H
        C FIND NEW HBAR FOR SECOND FIT
288     10 HBAR=(-FK(1)*C**2+(C**2-A**2)*FK(2)+A**2*FK(3))/(2.D 00*(-C*FK(1)
           1+(2)+A*FK(3)))
        C FIND NEW ESTIMATE OF MINIMUM
289        DO 15 J=1,N
290     15 XQ2(J)=X(2,J)+HBAR*ETA(J)
291        XQDM=0.D 00
292        DO 16 J=1,N
293     16 XQDM=(XQ(J)-XQ2(J))**2 + XQDM
294        XQDM=DSQRT(XQDM)
        C CHECK IF ANY IMPROVEMENT
295        IF(XQDM.LT.EPS)GO TO 17
296        DEL2=XQDM
297        FXQ2=F(XQ2)
        C CHECK IF FXQ2 BEST VALUE
298        IF(FXQ2.GT.FXQ)GO TO 18
299        IF(FXQ2.GT.FXB)GO TO 19
300        PRINT103,DEL1,DEL2
301        PRINT106,FXQ2,(XQ2(J),J=1,N)
302        GO TO 199
303     17 DEL2=EPS
304     18 IF(FXQ.GT.FXB)GO TO 19
305        PRINT104,DEL1,DEL2
306        PRINT106,FXQ,(XQ(J),J=1,N)
307        GO TO 198
308     19 PRINT105,DEL1,DEL2
309        PRINT106,FXB,(XB(J),J=1,N)
```

```
310          DO 153 J=1,N
311      153 XR(J)=XB(J)
312          GO TO 99
313      197 DO 150 J=1,N
314      150 XR(J)=X(2,J)
315          GO TO 99
316      198 DO 151 J=1,N
317      151 XR(J)=XQ(J)
318          GO TO 99
319      199 DO 152 J=1,N
320      152 XR(J)=XQ2(J)
321       99 RETURN
322      100 FORMAT(1HO,'F(X) NEARLY   LINEAR FOR FIRST FIT')
323      101 FORMAT(1HO,'QUADRATIC CHANGED ESTIMATE BY LESS THAN ',D14.6)
324      102 FORMAT(1HO,'F(X) NEARLY LINEAR FOR SECOND FIT'/' SHIFT ON FIRST F
             1IT WAS ',D14.6)
325      103 FORMAT(1HO,'BEST VALUE FROM SECOND QUADRATIC FIT'/'SHIFT ON FIRST
             1FIT WAS ',D14.6/' SHIFT ON SECOND WAS ',D13.6)
326      104 FORMAT(1HO,' NO IMPROVEMENT ON SECOND QUADRATIC FIT-USE FIRST'/'
             1SHIFT ON FIRST FIT WAS ',D14.6/' SHIFT ON SECOND WAS LESS THAN ',
             1D13.6)
327      105 FORMAT(1HO,' NO IMPROVEMENT BY QUADRATIC FITS'/' SHIFT ON FIRST F
             1IT WAS ',D14.6/' SHIFT ON SECOND FIT WAS ',D13.6)
328      106 FORMAT(1HO,'MINIMUM FX = ',D14.7,' AT X=',6D14.6)
329          END
```

REFERENCES

1. T. Alfrey, Jr., J. J. Bohrer, and H. Mark, "Copolymerization," Interscience, New York, 1952.

2. "Copolymerizatiom" (G. E. Ham, ed.), Interscience, New York, 1964.

3. C. Walling, "Free Radicals in Solution," John Wiley and Sons, Inc., New York, 1957.

4. F. R. Mayo and F. M. Lewis, *J. Amer. Chem. Soc., 66,* 1594 (1944).

5. T. Alfrey, Jr. and G. Goldfinger, *J. Chem. Phys., 12,* 322 (1944).

6. C. Walling and E. R. Briggs, *J. Amer. Chem. Soc., 67,* 1774 (1945).

7. R. G. Fordyce, E. C. Chapin, and G. E. Ham, *J. Amer. Chem., Soc., 70,* 2489 (1948).

8. C. Simeonescu, N. Asandie, and A. Liga, *Makromol. Chemie, 110,* 278 (1967).

9. A. Rudin, S. S. M. Chiang, H. K. Johnston, and P. D. Paulin, *Can. J. Chem., 50,* 1757 (1972).

10. P. W. Tidwell and G. A. Mortimer, *J. Macromol. Sci.-Revs., C4,* 281 (1970).

11. S. L. Aggarwal and F. A. Long, *J. Polym. Sci., 11,* 127 (1953).

12. K. Takemoto, Y. Kikuchi, and M. Imoto, *Chem. High Polym., (Tokyo), 23,* 459 (1966).

13. C. H. Bamford and S. Brumby, *Makromol. Chemie, 105,* 122 (1967).

14. G. G. Cameron and G. F. Esslemont, *Polymer (London), 13,* 435 (1972).

15. G. S. Franco and A. Leoni, *Polymer (London), 14,* 2 (1973).

16. M. Fineman and S. D. Ross, *J. Polym. Sci., 5,* 269 (1950).

17. P. W. Tidwell and G. A. Mortimer, *J. Polym. Sci. A., 3,* 369 (1965).

18. D. W. Behnken, *J. Polym. Sci.A , 2,* 645 (1964).

19. A. I. Yezrielev, E. L. Brokhina, and Y. S. Roskin, *Vysokomolekul. Soedin., A11,* 1670 (1969).

20. R. M. Joshi, *J. Macromol. Sci.-Chem., A7,* 1231 (1973).

21. R. M. Joshi and S. G. Joshi, *J. Macromol. Sci.-Chem., A5,* 1329 (1971).

22. H. K. Johnston and A. Rudin, *Macromolecules, 4,* 661 (1971).

23. D. Braun, W. Brendlein, and G. Mott, *Europ. Polym. J.*, *9*, 1007 (1973).

24. I. Skeist, *J. Amer. Chem. Soc.*, *68*, 1781 (1946).

25. H. O. Hartley, *Technometrics*, *3*, 269 (1971).

26. H. K. Johnston and A. Rudin, *J. Paint Technol.*, *42* (547), 429 (1970).

27. A. Rudin, W. R. Ableson, S. S. M. Chiang, and G. W. Bennett, *J. Macromol. Sci.-Chem.*, *A7*, 1203 (1973).

28. R. H. Wiley, S. P. Rao, J.-I. Jin, and K. S. Kim, *J. Macromol. Sci.-Chem.*, *A4*, 1453 (1970).

29. D. R. Montgomery and C. E. Fry, *J. Polym. Sci. C*, (25), 59 (1968).

30. H. J. Harwood, N. W. Johnston, and H. Pietrowski, *J. Polym. Sci. C*, (25) 23 (1968).

31. A. Rudin and R. G. Yule, *J. Polym. Sci. A-1*, *9*, 3009 (1971).

32. T. Alfrey, Jr. and C. C. Price, *J. Polym. Sci.*, *2*, 101 (1947).

33. T. Alfrey, Jr. and L. J. Young, "Copolymerization," (G. E. Ham, ed.), Chap. 11, Interscience, New York, 1964.

Chapter 5

THE USE OF COMPUTERS TO STUDY THE PENULTIMATE AND
THE CHARGE-TRANSFER POLYMERIZATION MODELS

Charles U. Pittman, Jr.
Thane D'Arcy Rounsefell

Department of Chemistry
The University of Alabama
University, Alabama

I. INTRODUCTION

To chemists and engineers studying the synthesis and manufac-
ture of copolymers, the copolymer equation, Eq. (1), has long been
the foundation by which the specific composition of a copolymer
could be designed [1,2]. This equation assumes (1) the copolymerization propagation process involves only the four equations, Eqs.
(2) to (5), (2) the penultimate monomer unit does not affect the
rate constants, (3) the rate constants are independent of the length
of the growing chain, and (4) the steady-state approximation holds.

$$\frac{d(M_1)}{d(M_2)} = \frac{M_1(r_1M_1 + M_2)}{M_2(r_2M_2 + M_1)} \tag{1}$$

where $r_1 = k_{11}/k_{12}$ and $r_2 = k_{22}/k_{21}$.

$$\text{\small www } M_1\cdot + M_1 \xrightarrow{k_{11}} \text{\small www } M_1 \text{—} M_1\cdot \tag{2}$$

$$\text{\small www } M_1\cdot + M_2 \xrightarrow{k_{12}} \text{\small www } M_1 \text{—} M_2\cdot \tag{3}$$

$$\text{\small www } M_2\cdot + M_1 \xrightarrow{k_{21}} \text{\small www } M_2 \text{—} M_1\cdot \tag{4}$$

$$\text{\small www } M_2\cdot + M_2 \xrightarrow{k_{22}} \text{\small www } M_2 \text{—} M_2\cdot \tag{5}$$

Many methods have been used to obtain r_1 and r_2 from copolymerization composition-conversion data. These include the intersection method [3], the curve-fitting method [3], and the linearization method (i.e., the Finman-Ross technique [4]) to treat low
conversion copolymerizations using the differential form of the copolymer equation, Eq. (1). Alternately, copolymerizations to higher
conversions have employed similar methods to fit the integrated form
of this equation [3,5]. Montgomery and Fry [6] pointed out the potential dangers of using the differential form of this equation.
They provided a computer program (modified FORTRAN IV) to accept
composition conversion data at different M_1^o/M_2^o ratios and conver-
age on r_1 and r_2 at any values of conversion. However, the method

they advocated suffered because they failed to apply a nonlinear least-squares method and, simultaneously, define the optimum starting M_1^o/M_2^o ratios to use in the experiments providing the data for r_1 and r_2 calculations. Despite these difficulties a huge number of r_1 and r_2 determinations have been reported and summarized in the literature [1,2] using all of the above techniques.

The great importance of using optimum experimental M_1^o/M_2^o ratios has been discussed in great detail by Behnken [7]. The great need to treat the problem in this manner, and the mathematical criteria for selecting these conditions has as its basis the work of Box and Lucas [8]. The general problem requires optimum experimental design for the nonlinear terminal model's copolymer equation, and its nonlinear solution is discussed by Box and Lucas [8]. Furthermore, the pitfalls of linearization in such problems was documented by Behnken [7]. Thus, in order to rigorously handle just the "simple" copolymer equation, Eq. (1), let alone the complex penultimate and charge-transfer polymerization models, one must (1) avoid linearization, and (2) use optimum experimental M_1^o/M_2^o ratios that minimize the area of the confidence region based on least-squares analysis.

A comprehensive discussion of the confidence region problem is given by Beal [9]. To minimize the confidence interval, it is necessary to choose conditions which make the determinant of the variance covariance matrix [8] as small as possible. All of these requirements were largely overlooked in the treatment of copolymer data until Tidwell and Mortimer developed a technique which applied these mathematical criteria for determining r_1 and r_2 from composition conversion data [10]. Tidwell and Mortimer have critically reviewed their computerized method and its application for obtaining r_1 and r_2, via optimum experimental M_1^o/M_2^o ratios, for the terminal model [11]. It is in their method that the basis for our approach to analyzing the penultimate and charge-transfer models using composition conversion data lies. The reader is encouraged to study Ref. 11 in conjunction with reading this chapter.

II. THE PENULTIMATE COPOLYMERIZATION MODEL
AND APPLICABLE FORTRAN IV PROGRAMS

The penultimate model of vinyl copolymerization, first introduced by Merz et al. [12], proposes that the next-to-last monomer
residue (i.e., the penultimate unit), as well as the terminal unit,
can influence the addition of the next monomer molecule to the
growing chain. Overall, it is possible that a maximum of eight
equations and four reactivity ratios can contribute to the propagation path. These are given below [Eqs. (6-13)]. In its most general form all eight equations would operate. However, this, probably

$$\sim\!\!\sim M_1 \text{—} M_1\cdot + M_1 \xrightarrow{k_{111}} \sim\!\!\sim M_1 \text{—} M_1 \text{—} M_1\cdot \tag{6}$$

$$\sim\!\!\sim M_1 \text{—} M_1\cdot + M_2 \xrightarrow{k_{112}} \sim\!\!\sim M_1 \text{—} M_1 \text{—} M_2\cdot \tag{7}$$

$$\sim\!\!\sim M_2 \text{—} M_1\cdot + M_1 \xrightarrow{k_{211}} \sim\!\!\sim M_2 \text{—} M_1 \text{—} M_1\cdot \tag{8}$$

$$\sim\!\!\sim M_2 \text{—} M_1\cdot + M_2 \xrightarrow{k_{212}} \sim\!\!\sim M_2 \text{—} M_1 \text{—} M_2\cdot \tag{9}$$

$$\sim\!\!\sim M_1 \text{—} M_2\cdot + M_1 \xrightarrow{k_{121}} \sim\!\!\sim M_1 \text{—} M_2 \text{—} M_1\cdot \tag{10}$$

$$\sim\!\!\sim M_1 \text{—} M_2\cdot + M_2 \xrightarrow{k_{122}} \sim\!\!\sim M_1 \text{—} M_2 \text{—} M_2\cdot \tag{11}$$

$$\sim\!\!\sim M_2 \text{—} M_2\cdot + M_1 \xrightarrow{k_{221}} \sim\!\!\sim M_2 \text{—} M_2 \text{—} M_1\cdot \tag{12}$$

$$\sim\!\!\sim M_2 \text{—} M_2\cdot + M_2 \xrightarrow{k_{222}} \sim\!\!\sim M_2 \text{—} M_2 \text{—} M_2\cdot \tag{13}$$

where

$$r_1 = \frac{k_{111}}{k_{112}} \qquad\qquad r_1' = \frac{k_{211}}{k_{212}}$$

$$r_2 = \frac{k_{222}}{k_{221}} \qquad\qquad r_2' = \frac{k_{122}}{k_{121}}$$

would be a rare case. If M_1 was sterically very bulky, k_{111} could be smaller than k_{211}. However, in such a situation k_{122} might equal k_{222} and/or k_{121} might equal k_{221}. Thus, the problem would be simplified to three parameters r_1, r_2, and r_1' since $r_2 = r_2'$. Similarly, electronic penultimate effects might be involved in which all eight equations are not unique. Furthermore, many cases exist where $\wedge\wedge\wedge M_2^\bullet + M_2$ does not proceed (that is, $k_{22} = 0$). Again the analysis is simplified.

Previously, the penultimate model approach has been extensively used to explain composition-conversion and sequence-distribution data which did not fit the terminal model [13-18]. Usually, these examples involved one of the simplifications outlined above. However, in some cases where composition-conversion data was satisfactorily modeled by the penultimate equation [see Eq. (14)], the penultimate model predicted very poorly when tested severely by monomer sequence analysis [18]. Berger and Kuntz [19] pointed out that the distinction between terminal and penultimate models, based on composition-conversion data, is frequently very difficult to obtain and that sequence-distribution studies are preferred. In addition, copolymerizations involving an electron-attracting and an electron-donating monomer could proceed through charge-transfer complexes [20-26].

In attempting to define the correct model by which a copolymerization takes place, the choice among terminal, penultimate, and charge-transfer models must involve extremely careful analysis. If the monomers have similar e values or if the copolymer compositions (or reactivity ratios) are insensitive to dilution, one may usually discard the charge-transfer mechanism [27]. A choice between penultimate and terminal models may proceed either by obtaining detailed sequence-distribution information [28,29] or "suitable" composition-conversion data. The sequence-distribution approach is clearly more sensitive in distinguishing these models [19], but it frequently is far more difficult experimentally to obtain sequence-distribution data. Thus, it remains important that improved general

methods be developed for using composition-conversion data to analyze the penultimate model. Furthermore, by choosing experiments
with the optimum design, the experimenter can obtain better values
of penultimate reactivity ratios which permit an objective analysis
of their reliability for a given experimental accuracy.

The general differential form of the penultimate copolymer
equation, Eq. (14), derived from Eqs. (6 to 13) using the steady-
state approximation [12], is given below. We have developed three

$$\frac{d(M_1)}{d(M_2)} = \frac{m_1}{m_2} = \frac{1 + r_1'\left(\dfrac{M_1}{M_2}\right)\left(\dfrac{M_2 + r_1 M_1}{M_2 + r_1' M_1}\right)}{1 + r_2'\left(\dfrac{M_2}{M_1}\right)\left(\dfrac{M_1 + r_2 M_2}{M_1 + r_2' M_2}\right)} \qquad \text{(Penultimate copolymer equation)} \qquad (14)$$

computer programs, the first of which permits the calculation of
the penultimate reactivity ratios r_1, r_1', r_2, and r_2', for Eq. (14)
without constraints. For example, this program can be used for data
obtained at any per cent conversion to polymer. The method is not
limited to special cases where $r_2 = r_2' = 0$, as in many prior treatments. The only input data needed are the initial mole fractions of
each monomer, the per cent conversion to polymer, the mole fractions
of each monomer in the polymer, and the molecular weights of each
monomer.

A second program has been written which finds the four optimum
initial values M_1^o/M_2^o to use in the copolymerization experiments
(which are applied in the first program). Approximations or available estimates of the four reactivity ratios are first needed. However, these two programs may be effectively used sequentially, in the
manner advocated by Behnken [7] for the terminal model, to allow
the most rapid approach toward obtaining these reactivity ratios.
Finally, a third program has been written which calculates the copolymer composition, at any conversion, using the penultimate
equation, Eq. (14). The input data for this program include the
four reactivity ratios, the molecular weights of the monomers, and
the M_1^o/M_2^o feed ratio. This program employs a numerical integration with a variable step size. The results can be printed for each

iteration or any multiple iteration. It permits the experimenter
to calculate the feed ratios and conversions necessary to construct
copolymers of any desired composition once the penultimate reactivity ratios are known. All three programs were written in FORTRAN IV
and have been run on an IBM 360/50 G level compiler. The execution
time for program I is largely dependent on the input data and the
actual values of the reactivity ratios, as well as the integration
increment selected. A printout of source decks, sample input and
output, and instructions will be made available on request. The
combination of approaches used in the first two programs provides
an improved approach to obtaining the penultimate parameters.

III. PENULTIMATE REACTIVITY RATIO PROGRAM I

Letting f stand for the right side of Eq. (14) we have

$$\frac{m_1}{m_2} = f$$

and since $m_1 = 1 - m_2$, then

$$m_2 = \frac{1}{1 + f} \tag{15}$$

The program starts with an initial estimate of the penultimate reactivity ratios, and it converges them to the best nonlinear least-
squares fit, Mardquart's algorithm [30,31] being used to speed convergence. The residuals for the sum of squares are obtained by
evaluating Eq. (15) for each data point and subtracting the experimental value of m_2. The program can be used at any conversion. To
allow for a drift in monomer ratios due to conversion, the program
integrates Eq. (15) stepwise by using the trapazoid rule. The step
size of this integration is chosen by the user. On each iteration
a check is made to see if the weight of the monomer removed from
M_1^o and M_2^o equals the weight of the polymer. When this happens
the calculation is ended, and $m_2 = (M_2^o - M_2)/(M_2^o - M_2 + M_1^o - M_1)$.
If M_1 or M_2 is reduced to zero during the integration the calculations stop, and m_2 is calculated directly without need of any integration.

The partial derivatives of Eq. (15) with respect to each of the reactivity ratios are needed for Marquardt's algorithm. These are determined numerically by finite differences with the above integration technique.

IV.   OPTIMUM EXPERIMENTAL M_1^o/M_2^o RATIOS FOR THE PENULTIMATE MODEL, PROGRAM II

The criterion to be used in selecting optimal experimental design is that of finding those initial feed ratios that minimize the area of the confidence region based on least-squares estimates [7-11]. To minimize the confidence interval, it is necessary to choose conditions which make the determinant of the variance covariance matrix [8] [this matrix is discussed in Ref. 8 and should not be confused with determinant (16)] as small as possible. Following Tidwell and Mortimer [10], we have applied this approach for obtaining optimum experimental M_1^o/M_2^o ratios in the penultimate model.

For the penultimate case, the determinant (16) was set up where $(\partial m_2/\partial r_1)(M_{2-1})$ is the partial derivative of Eq. (15) with respect to r_1 evaluated for M_2. The notation M_{2-1}, M_{2-2}, M_{2-3}, and M_{2-4}

$$\begin{vmatrix} \frac{\partial m_2}{\partial r_1}M_{2-1} & \frac{\partial m_2}{\partial r_1'}M_{2-1} & \frac{\partial m_2}{\partial r_2}M_{2-1} & \frac{\partial m_2}{\partial r_2'}M_{2-1} \\[2ex] \frac{\partial m_2}{\partial r_1}M_{2-2} & \frac{\partial m_2}{\partial r_1'}M_{2-2} & \frac{\partial m_2}{\partial r_2}M_{2-2} & \frac{\partial m_2}{\partial r_2'}M_{2-2} \\[2ex] \frac{\partial m_2}{\partial r_1}M_{2-3} & \frac{\partial m_2}{\partial r_1'}M_{2-3} & \frac{\partial m_2}{\partial r_2}M_{2-3} & \frac{\partial m_2}{\partial r_2'}M_{2-3} \\[2ex] \frac{\partial m_2}{\partial r_1}M_{2-4} & \frac{\partial m_2}{\partial r_1'}M_{2-4} & \frac{\partial m_2}{\partial r_2}M_{2-4} & \frac{\partial m_2}{\partial r_2'}M_{2-4} \end{vmatrix} \quad (16)$$

represents the four different optimized M_2 feed concentrations. Initial estimates of M_{2-1}, M_{2-2}, M_{2-3}, and M_{2-4} and the values of r_1, r_1', r_2, and r_2', for which optimum conditions are desired, constitute the input to the program. The program systematically varies

M_{2-1}, M_{2-2}, M_{2-3}, and M_{2-4} until it finds the four values which maximize the absolute value of the above determinant. Those four values are the optimum initial M_2 values to use in determining the penultimate reactivity ratios.

It is necessary to have estimates of the reactivity ratios prior to the calculation of reasonable M_1^o/M_2^o ratios for use in experiments. This can be accomplished by experience and/or by running a set of four to ten initial experiments and then computing crude reactivity ratios using the first program. Then these first reactivity ratio estimates are used in the second program to generate better initial monomer ratios for subsequent experiments. This approach is then repeated sequentially.

V. DISCUSSION OF PROGRAMS I TO III

On using the penultimate reactivity-ratio program I, reactivity ratios which give the best fit to experimental composition-conversion data can be found for any copolymer system. While r_1' and r_2' are not as sensitive to these experimental data as they are to sequence-distribution data, composition-conversion data are frequently easier to obtain; therefore, this program will have continuing utility.

It is possible for program I to converge on more than one minimum set of reactivity ratios depending on the experimental data and on the initial estimates of r_1, r_1', r_2, and r_2' used. This is due to the existence of local minima present in the four-dimensional coordinate space. To be sure that the correct reactivity ratios are found with a minimum of effort, the following technique is useful. First, assume a terminal model and calculate r_1 and r_2. These values of r_1 and r_2 are then used as initial estimates in the input data to program I. Furthermore, rather than feeding a single estimate of r_1' and r_2' into the program, a grid of r_1' and r_2' estimates is fed in. Most of these combinations will converge on the correct final computed values of the reactivity ratios, but some estimates will converge on local minima. The "correct" values are those

values of the reactivity ratios with the lowest sum of squares.
Program I has been specifically designed so the grid procedure may
be easily followed. The grid approach is also signigicant because
it can alert the user to experimental data for which there is no
unique fit. This is frequently a problem in fitting experimental
data to equations with many parameters.

This program has been extensively tested with 28 sets of test
data covering a wide range of penultimate reactivity-ratio combinations. In one type of test, "perfect" composition-conversion data
was generated at five initial mole fractions (0.1, 0.3, 0.5, 0.7,
and 0.9) of M_2 by using values of m_2 calculated from the composition-conversion program III at a given set of penultimate reactivity ratios. These composition-conversion data were then fed into
program I, and the reactivity ratios were calculated without difficulty by using the grid technique for r_1' and r_2' estimates and initial estimates of r_1 and r_2 from the terminal model. The use of
this program with experimental data will give "best" values of the
penultimate ratios. These "best" values can be made considerably
more reliable for a given experimental situation if experiments are
conducted at, or close to, the optimum starting concentrations. It
should be emphasized that in order to choose among terminal, penultimate, or charge-transfer models, accurate experimental data are
still required.

The use of this optimized, nonlinear least-squares sequential
approach to obtain penultimate ratios does not remove the need for
precise experiments. However, even given quite accurate experimental data, experiments performed at poorly chosen M_1^o/M_2^o ratios can
lead to such large confidence limits that the experiments are essentially wasted [7,10,11]. The use of linear estimation procedures can lead to errors which are often not considered by the experimenter [7,9,11]. On the other hand, our approach, like that
employed by Tidwell and Mortimer in the terminal model [10], guar-
antees the experimenter the efficient use of his data and allows an
objective probabilistic statement to be made regarding the reliability of his estimates. The experimenter can now choose his

experiments in the most economical way while simultaneously providing the most information per experiment.

When experimental data for a copolymerization, which proceeds
by the terminal model, is fed to program I, four reactivity ratios
will be obtained, and $r_1 = r_1'$ and $r_2 = r_2'$. Thus, the ability of
this technique to distinguish between the terminal and penultimate
models rests, in part, with the accuracy of the experimental data
and also with the user's choice of what really constitutes a penultimate effect. We have tested this effect by taking "perfect"
composition-conversion data (calculated by program III using opti-
mium M_1^o/M_2^o ratios from program II for a given set of reactivity
ratios r_1 and r_2, where $r_1 = r_1'$ and $r_2 = r_2'$) and perturbing these
data to simulate various types of experimental error. Some example
results are summarized in Table 1. The "perfect" experimental
composition-conversion data are given in columns 2 and 3. Columns
4 and 5 illustrate two types of error. The values of r_1, r_1' and r_2,
r_2' are compared below the table. Clearly, the errors introduced in
Table 1 result in only small perturbations of the relative reactivity ratios.

Similarly, experimental error has been introduced to a set of
"perfect" penultimate model composition-conversion data in Table 2.
Again, the perturbed data are given in columns 4 and 5, and calculated penultimate reactivity ratios are given below the table.
These ratios may be compared to the correct values used in the input. It is quite clear from Table 2 and its footnotes that, despite
experimental error, the penultimate effect is clearly portrayed.
Tables 1 and 2 also serve as sample input-output data against which
other users can test programs. For example, program I can be used
to calculate the penultimate ratios from three sets of composition-
conversion data in Table 2, columns 2, 4, and 5). Program III can
be tested by entering the given reactivity ratios and the initial
M_1^o/M_2^o feed ratios (column 1) and calculating the values of m_2 in
column 2 at the conversion listed in column 3.

The converging technique used in program II is a systematic
search constrained from 0 to 1 (mole fraction). When used

TABLE 1

Input and Output Results for the Terminal Model
for Both "Perfect" and Perturbed Data for Programs I, II, and III[a]

1 Optimum initial M_2^o	2 m_2 in polymer[b]	3 fractional conversion	4 m_2 in polymer[c,d]	5 m_2 in polymer[e,f]
0.0397	0.14236	0.05	0.13	0.15236
0.0397	0.13427	0.10	0.14	0.14427
0.0397	0.12619	0.15	0.11	0.13619
0.0397	0.11817	0.20	0.12	0.12817
0.1832	0.37894	0.05	0.36	0.38894
0.1832	0.36759	0.15	0.37	0.37759
0.5274	0.61480	0.05	0.60	0.62480
0.5274	0.61191	0.15	0.62	0.62191
0.8603	0.85183	0.05	0.84	0.86183
0.8603	0.85226	0.15	0.86	0.86226

[a]Molecular weights: $M_1 = 50$, $M_2 = 50$; relative reactivity ratios given: $r_1 = r_1' = 0.2$; $r_2 = r_2' = 0.8$.

[b]Calculated from program III by using given reactivity ratios.

[c]Values from column 2 have been truncated and 0.01 added or subtracted alternately.

[d]For these composition data and conversion data of column 3, the reactivity ratios become: $r_1 = 0.2139$, $r_1' = 0.2216$, $r_2 = 0.7788$, and $r_2' = 0.8164$.

[e]Values from column 2 have had 0.01 added to each.

[f]For these composition data and conversion data of column 3, the reactivity ratios become: $r_1 = 0.1740$, $r_1' = 0.1979$, $r_2 = 0.8928$, and $r_2' = 0.8460$.

TABLE 2

Input and Output Results for the Penultimate Model
for Both "Perfect" and Perturbed Data for Programs I, II, and III[a]

1 Optimum initial M_2^o	2 m_2 in polymer[b]	3 Fractional conversion	4 m_2 in polymer[c,d]	5 m_2 in polymer[e,f]
0.0440	0.14129	0.05	0.13	0.15129
0.0440	0.13507	0.10	0.14	0.14507
0.0440	0.12866	0.15	0.11	0.13866
0.0440	0.12210	0.20	0.13	0.13210
0.2351	0.36690	0.05	0.35	0.37690
0.2351	0.36131	0.15	0.37	0.37131
0.5822	0.60781	0.05	0.59	0.61761
0.5822	0.60677	0.15	0.61	0.61677
0.8732	0.85737	0.05	0.84	0.86737
0.8732	0.85826	0.15	0.86	0.86826

[a]Molecular weights: $M_1 = 50$, $M_2 = 50$; relative reactivity ratios given: $r_1 = 0.2$, $r_1' = 0.7$, $r_2 = 0.8$, $r_2' = 0.85$.

[b]Calculated from program III by using given reactivity ratios.

[c]Values in column 2 have been truncated and 0.01 added or subtracted alternately.

[d]For these composition data and conversion data of column 3, the reactivity ratios calculated by program I become: $r_1 = 0.2145$, $r_1' = 0.6678$, $r_2 = 0.7311$, and $r_2' = 0.7992$.

[e]Values in column 2 have had 0.01 added to each.

[f]For these composition data and conversion data of column 3, the reactivity ratios become: $r_1 = 0.1751$, $r_1' = 0.5882$, $r_2 = 0.9051$, and $r_2' = 0.7623$.

sequentially with a suitable grid of estimates in program I, the problem of systematically converging on a "false" minima is avoided.

Significant advances have been made in both the experimental determination [15,16] of and the convenient calculation [28,29] of sequence distributions. However, many sequence-distribution determinations are not experimentally accessible even with such powerful probes as 300 MHz PMR or ^{13}C nmr spectroscopy. Thus, the convenient computer handling of composition conversion information in the penultimate model using an optimized, nonlinear least-squares approach should prove useful.

VI. THE CHARGE-TRANSFER COPOLYMERIZATION MODEL

In addition to the penultimate effect, there exists another mechanism which can complicate normal analyses of copolymerizations by the classic terminal model. This involves the intervention of charge-transfer complexes in the propagation sequence [20-26]. This situation is most likely to occur when the two comonomers, M_1 and M_2, vary widely in electron affinity or in their e values (from the Q-e scheme) [1-3]. In the case where one monomer is strongly electron-releasing and the other strongly electron withdrawing, the two monomers can be in equilibrium with a charge-transfer complex, $M_1 \cdot M_2$, which will be designated here as CT.

$$M_1 + M_2 \rightleftarrows CT \tag{17}$$

$$K_{CT} = \frac{[CT]}{[M_1][M_2]} \tag{18}$$

The charge-transfer complex interaction has been described in various ways in the literature [32,33]. For our purposes, it will be sufficient to portray this interaction as an interaction of the highest occupied molecular orbital (HOMO) of the electron-rich monomer (i.e., the donor) with the lowest unoccupied molecular orbital (LUMO) of the electron poor monomer (i.e., the acceptor). In order for this interaction to be important, a number of symmetry conditions must be met. In simplified terms, the energy diagram below portrays a symmetry allowed charge-transfer interaction. This is a ground-state interaction and should not be confused in photo-induced charge-transfer complexes which are excited-state species.

(19)

The infrared spectra of such complexes are generally the sum of the spectra of M_1 and M_2 unless the interaction is very strong such as in $(CH_3)_3N \rightarrow I_2$ [34]. However, the ultraviolet-visible spectrum now exhibits a new, low energy transition called the charge-transfer band. By measuring the intensity of this transition it is frequently possible to determine K_{CT}, and a large number of these measurements have been made [32].

The formation of charge-transfer complexes in copolymerizations can be easily imagined between the following comonomer pairs where M_1 is the donor and M_2 the acceptor.

M_2 M_1 M_2 M_1
$e_2 = 2.1$ $e_1 = 0.50$ $c_2 = 2.25$ $e_1 = -1.37$

M_2
$e_2 = 1.96$

M_1
$e_1 = -1.11$

M_2
$e_2 = 1.96$

M_1
$e_1 = -2.0$

$CH_2 = CH - CN$

M_2
$e_2 = ?$

M_1
$e_1 = -1.40$

M_2
$e_2 = 1.20$

$CH_2 = C \begin{smallmatrix} CN \\ COCH_3 \\ O \end{smallmatrix}$

$CH_2 = CH - O - C - CH_3$
                   O

M_1
$e_1 = -1.9$

M_2
$e_2 = 2.10$

M_1
$e_1 = -0.22$

In general, when only the terminal model mechanism is operating, the larger the absolute value between e_2 and e_1 becomes, the smaller the product $r_1 \cdot r_2$ becomes. As $r_1 \cdot r_2$ approaches zero, the copolymer structure approaches the alternative structure $-M_1-M_2-M_1-M_2-$. Similarly, if the growing polymer chain reacts with a charge-transfer complex, one might expect an alternating copolymer structure, since as CT is incorporated into the polymer, both M_1 and M_2 are joined.

$$\text{\Large$\sim$} M_1 \cdot \ + \ CT \ \xrightarrow{\ k_{1c_2}\ } \ \text{\Large$\sim$} M_1 \text{-} M_2 \text{-} M_1 \cdot \tag{20}$$

$$\text{\Large$\sim$} M_2 \cdot \ + \ CT \ \xrightarrow{\ k_{2c_1}\ } \ \text{\Large$\sim$} M_2 \text{-} M_1 \text{-} M_2 \cdot \tag{21}$$

This, of course, assumes that $\text{\small$\sim$} M_1 \cdot$ prefers to add to the M_2 end of CT and that $\text{\small$\sim$} M_2 \cdot$ prefers to add to the M_1 end of CT. (We continue to define M_1 as the donor molecule and M_2 as the acceptor molecule here.)

As the $e_2 - e_1$ difference becomes larger, the polar stabilization of the transition states in the propagation reactions are influenced in the same way in the terminal model as in the charge-transfer model [see Eqs. (22) and (23)]. This is why it is so difficult to distinguish the charge-transfer process from the terminal model, especially if they are both simultaneously and competitively taking place in a given copolymerization.

$$\left[\text{\Large$\sim$} M_1 \cdot \overset{\delta+}{} \text{-----} \overset{\delta-}{M_2} \right]^{\dagger} \ \rightarrow \ \text{\Large$\sim$} M_1 \text{-} M_2 \cdot \quad \text{Terminal model} \tag{22}$$

$$M_1 = \text{donor}$$

$$\left[\text{\Large$\sim$} M_1 \cdot \overset{\delta+}{} \text{------} \overset{\delta-}{M_2} \leftarrow \overset{\delta+}{M_1} \right]^{\dagger} \ \rightarrow \ \text{\Large$\sim$} M_1 \text{-} M_2 \text{-} M_1 \cdot \quad \begin{array}{c} \text{Charge-transfer} \\ \text{model} \end{array} \tag{23}$$

In the most general case, eight propagation steps could be taking place in a copolymerization where both the terminal and the charge-transfer mechanisms are operating. They are summarized as Eqs. (24 to 31). The first four are those of the terminal model, and the last four involve the growing polymer chain adding to the CT complex. In order to maintain complete generality, the growing chain, terminated by either M_1 or M_2, is pictured as adding to both the M_1 and M_2 ends of the complex. This leads to the last four equations. It would be extremely unlikely that all processes (28 to 31) would operate significantly in any given copolymerization.

$$\sim\!\!\sim M_1\cdot + M_1 \xrightarrow{\ k_{11}\ } \sim\!\!\sim M_{\bar{1}}\!-\!M_1\cdot \qquad (24)$$

$$\sim\!\!\sim M_1\cdot + M_2 \xrightarrow{\ k_{12}\ } \sim\!\!\sim M_{\bar{1}}\!-\!M_2\cdot \qquad (25)$$

$$\sim\!\!\sim M_2\cdot + M_1 \xrightarrow{\ k_{21}\ } \sim\!\!\sim M_{\bar{2}}\!-\!M_1\cdot \qquad (26)$$

$$\sim\!\!\sim M_2\cdot + M_2 \xrightarrow{\ k_{22}\ } \sim\!\!\sim M_{\bar{2}}\!-\!M_2\cdot \qquad (27)$$

$$\sim\!\!\sim M_1\cdot + [M_1 \rightarrow M_2] \xrightarrow{\ k_{1c_1}\ } \sim\!\!\sim M_{\bar{1}}\!-\!M_{\bar{1}}\!-\!M_2\cdot \qquad (28)$$

$$\sim\!\!\sim M_1\cdot + [M_2 \leftarrow M_1] \xrightarrow{\ k_{1c_2}\ } \sim\!\!\sim M_{\bar{1}}\!-\!M_{\bar{2}}\!-\!M_1\cdot \qquad (29)$$

$$\sim\!\!\sim M_2\cdot + [M_1 \rightarrow M_2] \xrightarrow{\ k_{2c_1}\ } \sim\!\!\sim M_{\bar{2}}\!-\!M_{\bar{1}}\!-\!M_2\cdot \qquad (30)$$

$$\sim\!\!\sim M_2\cdot + [M_2 \leftarrow M_1] \xrightarrow{\ k_{2c_2}\ } \sim\!\!\sim M_{\bar{2}}\!-\!M_{\bar{2}}\!-\!M_1\cdot \qquad (31)$$

These equations portray the intervention of ground state complexes only, and our discussion will be restricted to this class of copolymerizations. Dilution studies are often used to see if a charge-transfer complex mechanism is operating [27]. The values of r_1 and r_2 derived, assuming the terminal model holds, are examined as a function of monomer concentration. If marked changes in r_1 and r_2 are observed and e_2-e_1 is large, the intervention of CT complexes is often assumed.

Seiner and Litt [35] have recently derived a kinetic model, based on Eqs. (24 to 31), to account for composition conversion data in copolymerizations where both terminal and CT mechanisms compete. In its most general form, it assumes the growing chain radical can add to either the uncomplexed monomer or to the CT complex. Using the usual assumptions of conditional probability and the steady-state approximation, the following general equation was derived [35].

$$y = \frac{m_1}{m_2} = \frac{\left\{1 + \left[\dfrac{r_2}{r_{2c2}} + \dfrac{r_2}{r_{2c1}}\right]\dfrac{[CT]}{[M_1]} + \left[\dfrac{r_1}{r_{1c2}}\dfrac{[CT]}{[M_2]}\right]\dfrac{1 + \dfrac{r_2}{r_{2c2}}\dfrac{[CT]}{[M_1]}}{1 + \dfrac{r_1}{r_{1c1}}\dfrac{[CT]}{[M_2]}}\right\}\left\{r_2\dfrac{[M_2]}{[M_1]} + 1 + \left[\dfrac{2r_2}{r_{2c2}} + \dfrac{r_2}{r_{2c1}}\right]\dfrac{[CT]}{[M_1]} + \left[\dfrac{r_1}{r_{1c2}}\dfrac{CT}{M_2}\right]\dfrac{1 + \dfrac{r_1}{r_{2c2}}\dfrac{[CT]}{[M_1]}}{1 + \dfrac{r_1}{r_{1c1}}\dfrac{[CT]}{[M_2]}}\right\}}{\left\{1 + \left[\dfrac{r_1}{r_{1c1}} + \dfrac{r_1}{r_{1c2}}\right]\dfrac{[CT]}{[M_2]} + \left[\dfrac{r_2}{r_{2c1}}\dfrac{[CT]}{[M_1]}\right]\dfrac{1 + \dfrac{r_1}{r_{1c1}}\dfrac{[CT]}{[M_2]}}{1 + \dfrac{r_2}{r_{2c2}}\dfrac{[CT]}{[M_1]}}\right\}\left\{r_1\dfrac{[M_1]}{[M_2]} + 1 + \left[\dfrac{2r_1}{r_{1c1}} + \dfrac{r_1}{r_{1c2}}\right]\dfrac{[CT]}{[M_2]} + \left[\dfrac{r_2}{r_{2c1}}\dfrac{[CT]}{[M_1]}\right]\dfrac{1 + \dfrac{r_1}{r_{1c1}}\dfrac{[CT]}{[M_2]}}{1 + \dfrac{r_2}{r_{2c2}}\dfrac{[CT]}{[M_1]}}\right\}}$$

(32)

where

$$r_1 = \frac{k_{11}}{k_{12}} \qquad r_{1c1} = \frac{k_{11}}{k_{1c1}} \qquad r_{1c2} = \frac{k_{11}}{k_{1c2}}$$

$$r_2 = \frac{k_{22}}{k_{21}} \qquad r_{2c2} = \frac{k_{22}}{k_{2c2}} \qquad r_{2c1} = \frac{k_{22}}{k_{2c1}}$$

y = the ratio of M_1/M_2 being incorporated, instantaneously, into the copolymer; M_1 and M_2 are the instantaneous concentrations of uncomplexed monomers in solution; and CT is the instantaneous concentration of the charge-transfer complex in solution.

As we saw earlier in dealing with the penultimate model, the eight equations lead to more than the two reactivity ratios found in the terminal model. They are r_1, r_2, r_{1c1}, r_{1c2}, r_{2c1}, and r_{2c2}, each of which we must know in order to evaluate the initial rate of incorporation of monomers, $d(M_1)/d(M_2)$, into the polymer. Furthermore, these values must be known if we are to use the integrated form of this equation to predict the total composition of the polymer at any percentage conversion (given, of course, the starting values of $[M_1^o]$, $[M_2^o]$, and $[CT^o]$). This equation is integrated numerically. It has not yet been solved analytically.

The $[CT]$ is related to $[M_1]$ and $[M_2]$ by

$$K_{CT} = \frac{CT}{[M_1][M_2]} = \frac{[CT]}{[M_1^o - CT][M_2^o - CT]} \tag{33}$$

where $[M_1^o]$ and $[M_2^o]$ are the total concentrations of M_1 and M_2 charged to the reaction initially, or if the reaction has already proceeded to some percent conversion, $[M_1^o]$ and $[M_2^o]$ in this key expression means the total concentrations of the complexed plus uncomplexed monomers.

If K_{CT} is small, as is very often the case in charge-transfer complexes since the interaction is usually a weak one, then

$$K_{CT} = \frac{[CT]}{[M_1^o][M_2^o]} \tag{34}$$

Seiner and Litt have examined literature data for the copolymerizations of several monomer pairs including styrene-β-cyanoacrolein, methyl acrylate-1,1-diphenylethylene, and vinyl acetate-hexafluoroacetone [35]. In these studies they have used this simplification as well as others. However, we find this an unnecessary simplification because $[M_1]$, $[M_2]$, and $[CT]$ can be readily calculated from $[M_1^o]$, $[M_2^o]$, and K_{CT} by rearranging Eq. (33) so that it can be solved by the quadratic formula. This is a time consuming task which lends itself to using the computer.

Equation (32) or its integrated form may, in theory, be used to determine the six reactivity ratios r_1, r_2, r_{1c1}, r_{1c2}, r_{2c1}, and r_{2c2}. In order to do this one must have composition-conversion data. If the differential form is used, polymerizations to very low conversions (<5% and preferably <1%) must be used. Using numerical integration of Eq. (32), any conversion can be employed, but yields below 15 to 20% are desired so that each experiment has a reasonably high weight value. Given sufficient composition-conversion data, it should be possible to determine these reactivity ratios by a nonlinear least-squares fitting technique. The fitting technique must be nonlinear to avoid the many mathematical pitfalls which plague the methods where linearization techniques are used. A desirable method is the one used in the solution of the penultimate effect described earlier in this chapter. The real problems facing such computations include (1) what defines a unique fit or does such "uniqueness" exist, (2) how does one avoid falling into false local minima, (3) how does one define the optimum experimental M_1^o/M_2^o ratios and concentrations, (4) can one define in any way the confidence intervals associated with the data?

Since the [CT] is a function of K_{CT}, the value of K_{CT} could be determined as another parameter in the nonlinear least-squares approach. Alternately K_{CT} could be determined experimentally by uv or possibly (in special cases) by nmr techniques [32]. K_{CT} in each equation has been considered to be a constant. However, in actual practice during a copolymerization this may not be so. Any consideration of K_{CT} in copolymerizations where the concentration of monomers are fairly high raises difficult questions. For example, incorporation of the monomers into the polymer has the effect of continuously varying bulk solvent properties such as dielectric constant. Thus, K_{CT} in such cases is a function of the percent conversion. Furthermore, changes in solvent properties can cause the extinction coefficient of the complex, ε, to change. Thus, if K_{CT} were actually decreasing as the solvent dielectric constant decreased during a copolymerization, an increase in ε could make the product $K\varepsilon$ appear to remain approximately unchanged. This process has been previously described under the name of contact charge-transfer [36,37].

Given all these complications it is not surprising that a rigorous solution to the general problem is elusive. To date, only a simple case of the general model has actually been employed. This is the case where $\wedge\wedge\wedge$ $M_2^{\bullet}$ + M_2 $-\!/\!/\!\!\rightarrow$; that is, where k_{22} = 0. When this occurs k_{2c2} (in all probability) is also 0. In this special (but reasonably common) case, Eq. (32) is reduced to Eq. (35).

$$y = \frac{m_1}{m_2} = 1 + \frac{\dfrac{1}{r_{1c1}} + \dfrac{[M_1]}{[CT]}}{\left(\dfrac{1}{r_{1c1}} + \dfrac{1}{r_{1c2}}\right) + \dfrac{M_2}{[CT]r_1}\left[1 + \left(\dfrac{[CT]r_2}{[M_1]r_{2c1}}\right)\left(1 + \dfrac{[CT]r_1}{[M_2]r_{1c1}}\right)\right]}$$

(35)

Allowing

$$\frac{1}{r_{1c}} = \frac{1}{r_{1c1}} + \frac{1}{r_{1c2}}$$

(36)

and

$$\theta = \left(\frac{[CT]}{[M_1]}\frac{r_2}{r_{2c1}}\right)\left(1 + \frac{[CT]}{[M_2]}\frac{r_1}{r_{1c1}}\right)$$

(37)

Equation (35) may now be written as (38).

$$y = 1 + \frac{\dfrac{1}{r_{1c1}} + \dfrac{[M_1]}{[CT]}}{\dfrac{1}{r_{1c}} + \dfrac{[M_2]}{[CT]r_1}[1 + \theta]}$$

(38)

When the product $[CT]$ r_2 $\ll$ $[M_1]$ r_{2c1}, then θ is very small and may be neglected. In cases where K_{CT} is small (that is, K_{CT} <0.03) this should hold since the value of $[CT]$ will be small. Thus Eq. (38) further reduces to (39) where another parameter has been eliminated.

$$y = 1 + \frac{\dfrac{1}{r_{1c1}} + \dfrac{[M_1]}{[CT]}}{\dfrac{1}{r_{1c}} + \dfrac{[M_2]}{[CT]\ r_1}}$$

(39)

In this version, only three reactivity ratios and K_{CT} remain to be determined. Thus, it is Eq. (39) which should be applied to systems where $\wedge\wedge\wedge M_2^{\bullet}$ does not add to M_2 and the value of K_{eg} is small.

Example copolymerizations where Eq. (39) might be employed include styrene- maleic anhydride, styrene- fumaronitrile, or vinyl-ferrocene-acrylonitrile. In fact, Seiner and Litt [35] have evaluated several sets of literature data using this approach. However, it must be most strongly emphasized that studies now in the literature were (1) not performed at the optimum starting concentrations of M_1 and M_2 (indeed for the charge-transfer model these concentrations have never even been rigorously defined), (2) not calculated using nonlinear least-squares methods which appropriately weighted each data point, and (3) not performed with sufficient data, including enough experiments at the appropriate concentrations, reproducibility studies, and studies testing the calculated ratios by performing a series of copolymerization experiments at alternate starting concentrations. We feel that when these criteria are met, and only then, should one begin to assign a mechanism involving charge-transfer complexes in competition with the terminal model for composition conversion data. The correct approach to this problem involves an optimization technique like that described for the penultimate effect applied to the models given in Eqs. (32), (38), and (39).

VII. CHARGE-TRANSFER COPOLYMERIZATION MODEL FORTRAN IV PROGRAMS

As with the penultimate model, three programs have been written for the charge-transfer copolymerization model. The first program (Program CT-1) permits the calculation of r_1, r_{1c1}, r_{1c2}, r_2, r_{2c2}, r_{2c1}, and K from composition-conversion data. The second program (Program CT-2) allows one to design the optimum concentrations of M_1^o and M_2^o to use in experimental work. This program may be used sequentially with the first program (and careful experimental work) to ultimately provide the best approach in obtaining r_1, r_{1c1}, r_{1c2}, r_2, r_{2c2}, r_{2c1}, and K. The third program (Program CT-3) calculates the copolymer composition, at any conversion, using charge-transfer model Eq. (32) in the most general case.

A. Program CT-1

This program is constructed similar to Program I for the penultimate model. Starting with initial estimates of the reactivity ratios $(r_1, r_{1c1}, r_{1c2}, r_2, r_{2c2}, r_{2c1})$ and K, it converges them to the best nonlinear least-squares fit using Marquardt's algorithm. It allows for multiple estimates of these parameters to avoid the possibility of falling into a false local minima. This "sector-cutting" technique is similar to that employed before in penultimate program I. However, there are a few differences in this program CT-1:

1. Instead of the one equation (14) in the penultimate program, this program can employ Eqs. (32), (38), or (39). The user selects the particular case he wishes to use.
2. The user may declare any combination of the parameters $(r_1, r_{1c1}, r_{1c2}, r_{2c2}, r_{21}, K)$ to be set. If a parameter is set, the program assumes it is correct and does not attempt to change it. This would be especially useful in a case where K had been determined by separate experiments or where either r_1 or r_2 were known.
3. The parameters are constrained to be greater than zero.

B. Program CT-2: Optimum Experimental Conditions

Again, the criterion to be used in selecting optimal experimental design is that of finding those initial feed ratios that minimize the area of the confidence region [7-11]. Thus, as was done for the penultimate model program II, determinants from Eqs. (32), (38), or (39) were constructed. The user selects the equation he wishes to use. For example, the determinant of Eq. (39) is listed as Eq. (40). The other determinants from Eqs. (32) and (38) have been left out of this discussion to save space.

Program CT-2 differs from the penultimate optimization program in the following ways: (1) the determinant is constructed from Eqs. (32), (38), or (39) instead of (14). The user selects the equation he wishes to use. (2) The user may declare any combination of the parameters to be set. If a parameter is set, the program assumes it is known and does not try to find optimum conditions for that parameter. (3) Since there are three different

$$
\begin{vmatrix}
\frac{\partial m_2}{\partial r_1}(M_{1-1},M_{2-1}) & \frac{\partial m_2}{\partial r_{1c1}}(M_{1-1},M_{2-1}) & \frac{\partial m_2}{\partial r_{1c2}}(M_{1-1},M_{2-1}) & \frac{\partial m_2}{\partial K}(M_{1-1},M_{2-1}) \\[3mm]
\frac{\partial m_2}{\partial r_1}(M_{1-2},M_{2-2}) & \frac{\partial m_2}{\partial r_{1c1}}(M_{1-2},M_{2-2}) & \frac{\partial m_2}{\partial r_{1c2}}(M_{1-2},M_{2-2}) & \frac{\partial m_2}{\partial K}(M_{1-2},M_{2-2}) \\[3mm]
\frac{\partial m_2}{\partial r_1}(M_{1-3},M_{2-3}) & \frac{\partial m_2}{\partial r_{1c1}}(M_{1-3},M_{2-3}) & \frac{\partial m_2}{\partial r_{1c2}}(M_{1-3},M_{2-3}) & \frac{\partial m_2}{\partial K}(M_{1-3},M_{2-3}) \\[3mm]
\frac{\partial m_2}{\partial r_1}(M_{1-4},M_{2-4}) & \frac{\partial m_2}{\partial r_{1c1}}(M_{1-4},M_{2-4}) & \frac{\partial m_2}{\partial r_{1c2}}(M_{1-4},M_{2-4}) & \frac{\partial m_2}{\partial K}(M_{1-4},M_{2-4})
\end{vmatrix}
$$

$$(40)$$

equations possible and since parameters can be set, the determinant
[which corresponds to (16) for the penultimate model] may vary in
size from 1 x 1 to 7 x 7. The determinant (40) for Eq. (39) is
4 x 4 for example. It should be noted that for the charge-transfer
model, it is necessary to find two independent variables (M_1 and
M_2) for each row of the determinant, <u>since M_1 and M_2 are concentrations and are independent of each other.</u> In the penultimate model,
only one independent variable M_2 was found for each row since M_1
and M_2 in that model were not concentrations but mole fractions
which are not independent of each other.

The determinant shown for Eq. (39) assumes that no parameters
are set. If a parameter is set, the column corresponding to that
parameter is eliminated. For instance, if K were set, the last
column would be eliminated. The number of rows is adjusted to
equal the number of columns. (4) The user may set any combination
of the independent variables (M_1 and M_2). When a variable is set,
the program will assume that variable to be an optimum starting
concentration and will not change it. The setting of one or more
of the variables becomes desirable when the determinant becomes
larger than about 4 x 4. This is because the computation time increases by a factor of nine for each additional column (row) of the
determinant, but decreases by a factor of three for each variable
that is set.

C. Program CT-3

This program is similar to Program III for the penultimate model. It uses Eqs. (32), (38), or (39) instead of (14). The user first selects the appropriate equation. The program calculates mole fraction of m_1 and m_2 in a copolymer at any conversion for the charge-transfer model given the parameters (r_1, r_{1c1}, r_{1c2}, r_2, r_{2c2}, r_{2c1}, K) and the initial concentrations of M_1 and M_2.

D. Discussion of CT-1, CT-2, and CT-3

The use of multiple estimates in calculating the reactivity ratios and the need for optimum experimental conditions has been previously discussed for the penultimate model. In the charge-transfer model it might be possible to determine one of the parameters, K, by independent experiments, although this would be an exceptional case. The programs are written to accept an independent value for K if the user desires.

Input-Output data are given in Tables 3 and 4. For the case where M_2 will not add to itself (that is, $k_{22} = 0$, $r_2 = 0$), Eq. (39) has been employed and one set of sample results are given in Table 3. Using Program CT-1, the composition-conversion data given in columns 1 to 4 were converged to give the parameters in footnote a. Very small perturbations in the polymer compositions did not disturb the ability to converge on a new set of parameters (see footnotes e and g of Table 3). These perturbations are smaller than experiments could detect. Thus, larger perturbations were made. Perturbations of 0.01 or 0.02 will converge on somewhat different values of the reactivity ratios and K. However, it is important to note that the size of acceptable perturbations depend on (1) if they are random and (2) the total number of experiments conducted. With random perturbations, larger experimental errors can be tolerated than with systematically biased errors. Given a sufficiently large number of experiments and random perturbations, fairly large experimental errors become acceptable. These are mutually interdependent factors.

Using the values of r_1, r_{1c1}, r_{1c2}, and K given in footnote a, the optimum starting concentrations were derived using Program CT-2. These are given in columns 1 and 2 of Table 3. Finally, using the same parameters and the fractional conversions in column 3, the polymer compositions in column 4 were calculated. It is possible to calculate all the parameters simultaneously. However, if the experimental data is not precise, the program may not converge, or it may converge on poor answers, since an error in K may be offset by compensating errors in the other parameters. The programs do not take into account any changes in K and/or the concentration of charge-transfer complex due to changes in the solvent properties which occur when the monomer concentrations change. Since optimum experimental conditions call for a wide range of monomer concentrations, this effect is a problem. A change in the programs to adjust for these effects could be easily made by anyone providing the proper algorithm to account for these effects. Currently, the best approach in handling K is viewing it as a parameter, and avoiding assigning it as the actual equilibrium constant at a particular set of conditions [35].

Next the completely general case was considered where neither r_1 nor r_2 was zero. In this case all seven parameters (r_1, r_{1c1}, r_{1c2}, r_2, r_{2c2}, r_{2c1}, and K) must be considered. Thus, Eq. (32) must be employed. In order to pick a system which might have some reality, the values of r_1 and r_2 were chosen based on calculations from the Q-e values of the monomer pair α-chloroacrylonitrile and ethyl vinyl ether. Then the values of r_{1c1}, r_{1c2}, r_{2c2}, and K (see footnote a in Table 4) were picked in a realistic range based on electronic and steric considerations. Thus, the values of the parameters used in program CT-2 were ones which could fall in a realistic range experimentally. Using CT-2 and the chosen values of the parameters, the optimum initial $[M_1^o]$ and $[M_2^o]$ values were calculated. (These are shown in columns 1 and 2 of Table 4.) Using the chosen parameters and these optimum initial starting concentrations, the polymer compositions at the fractional conversions

TABLE 3

Input and Output Results for the Charge-Transfer Model for Both "Perfect" and Perturbed Data for Programs CT-1, CT-2, and CT-3[a]

1 Optimum initial $[M_1^o]$[b]	2 Optimum initial $[M_2^o]$[b]	3 Fractional conversion	4 m_2 in polymer[c]	5 m_2 in polymer[d,e]	6 m_2 in polymer[f,g]
5.0	5.0	0.05	0.36107	0.3636	0.3586
5.0	5.0	0.10	0.36233	0.3648	0.3598
5.0	5.0	0.15	0.36367	0.3662	0.3612
5.0	5.0	0.20	0.36511	0.3676	0.3626
5.0	5.0	0.25	0.36665	0.3692	0.3641
0.114	0.106	0.05	0.34491	0.3474	0.3424
0.114	0.106	0.10	0.34651	0.3490	0.3440
0.114	0.106	0.15	0.34818	0.3507	0.3457
0.114	0.106	0.20	0.34994	0.3524	0.3474
0.114	0.106	0.25	0.35179	0.3543	0.3493
5.0	1.1	0.05	0.18131	0.1838	0.1788
5.0	1.1	0.10	0.18104	0.1835	0.1785
5.0	1.1	0.15	0.18077	0.1833	0.1782

(continued)

5.0	1.1	0.20	0.18051	0.1830	0.1780
5.0	1.1	0.25	0.18026	0.1828	0.1778
0.37	0.1	0.05	0.19558	0.1981	0.1931
0.37	0.1	0.10	0.19589	0.1984	0.1934
0.37	0.1	0.15	0.19622	0.1987	0.1937
0.37	0.1	0.25	0.19692	0.1994	0.1944
0.37	0.1	0.20	0.19656	0.1991	0.1941

[a] Molecular weights: $M_1 = 100$; $M_2 = 100$; Parameters: $r_1 = 0.85$, $r_{1c1} = 0.80$, $r_{1c2} = 0.30$, $K = 0.0075$; Eq. (39) used in calculations.

[b] Calculated with program CT-2.

[c] Calculated with program CT-3.

[d] Values from column 4 have been rounded and 0.0025 added.

[e] For these composition data and conversion data of column 3 the parameters calculated by CT-3 are: $r_1 = 0.83$, $r_{1c1} = 0.95$, $r_{1c2} = 0.42$, $K = 0.012$. If K is set at 0.0075 the results are: $r_1 = 0.83$, $r_{1c1} = 0.90$, $r_{1c2} = 0.28$.

[f] Values from column 4 have been rounded and 0.0025 subtracted.

[g] For the composition data of columns 5 and 6, and the conversion data of column 3 the parameters calculated by CT-3 are: $r_1 = 0.85$, $r_{1c1} = 0.86$, $r_{1c2} = 0.39$, $K = 0.010$. If K is set at 0.0075, the results are: $r_1 = 0.85$, $r_{1c1} = 0.80$, $r_{1c2} = 0.30$.

TABLE 4

Input and Output Results for the Completely General Charge-Transfer Model [Eq. (32)] Using both "Perfect" and Perturbed Data in Programs CT-1, CT-2, and CT-3[a]

1 Optimum initial M_1^o	2 Optimum initial M_2^o	3 Fractional conversion	4 m_2 in polymer[b]	5 m_2 in polymer[c,e]	6 m_2 in polymer[d]
5.0	5.0	0.05	0.31994	0.32994	0.30994
5.0	5.0	0.10	0.32073	0.33073	0.31073
0.10	5.0	0.05	0.60797	0.61797	0.59797
0.10	5.0	0.10	0.80363	0.81363	0.79363
5.0	1.09	0.05	0.16159	0.17159	0.15159
5.0	1.09	0.10	0.16076	0.17076	0.15076
0.113	0.10	0.05	0.26002	0.27002	0.25002
0.113	0.10	0.10	0.26280	0.27280	0.25280
1.5	2.34	0.05	0.33992	0.34992	0.32992
1.5	2.34	0.10	0.34247	0.35247	0.33247
0.404	5.0	0.05	0.47234	0.48234	0.46234
0.404	5.0	0.10	0.48059	0.49059	0.47059
0.1	2.47	0.05	0.50083	0.51083	0.49083
0.1	2.47	0.10	0.61089	0.62089	0.60089

[a] Molecular weights: $M_1 = 100$, $M_2 = 100$; Parameters: $r_1 = 1.7$, $r_{1c1} = 0.1$, $r_{1c2} = 0.3$, $r_2 = 0.0015$, $r_{2c2} = 0.2$, $r_{2c1} = 0.001$, $K = 0.01$. Equation (32) was used in the calculations.

[b] Calculated from Program CT-3 using parameters given in footnote a.

[c] Values from column 4 have had 0.01 added.

[d] Values from column 4 have had 0.01 subtracted.

[e] Using the Composition Data of columns 5 and 6, the following parameters were calculated by CT-1: $r_1 = 1.7$, $r_{1c1} = 0.1$, $r_{1c2} = 0.1$, $r_2 = 0.29$, $r_2 = 0.0015$, $r_{2c2} = 0.18$, $r_{2c1} = 0.0010$, $K = 0.0099$.

(listed in column 3) were calculated using CT-3. These values of m_2 are listed in column 4. Alternately, using the optimum starting concentrations and both given conversions and polymer compositions, CT-1 was employed to calculate the value of the reactivity parameters (see footnote a). Thus, any user can check these programs with this set of test data.

Next, the sensitivity of the programs to errors in polymer composition was examined. Columns 5 and 6 give a list of biased errors to the high and low side of m_2, respectively. Also alternating data points to the high and low side (balanced error) were examined. Biased data to the high side would not easily converge in a reasonable computer time. However, when the balanced errors were introduced, program CT-1 rapidly converged the data to the values for the parameters shown in footnote e. The 95% confidence limits for these calculations were substantial.

In practice, it was found expedient to use Mortimer-Tidwell terminal model program with raw composition-conversion data to get crude values of r_1 and r_2 first. These values serve as the estimates of r_1 and r_2 to feed into CT-1. Alternately, these values plus "sector cutting" can be used. This saves considerable time compared to the use of random guesses for r_1 and r_2. The measurement of fractional conversions can also be quite critical. Note, for example, the third and fourth horizontal rows of data in Table 4. At initial $[M_1^{\,o}]$ and $[M_2^{\,o}]$ of 0.1 and 5.0, respectively, the fraction of M_2 in polymer went from 0.608 to 0.804 as the percent conversion went from 5 to 10%. This emphasizes the need for careful experiments.

Using this rigorous approach demonstrated that no adequate composition-conversion data currently exists in the literature which can be used to test for a combined terminal-charge transfer polymerization model. Secondly, this technique provides the guide for what experiments should be performed and how many should be conducted. Third, the method requires, for most situations, highly accurate composition-conversion data. This necessary accuracy surpasses that which is normally obtainable. However, for certain combinations of reactivity parameters, this is less critical than in other regions. Thus, the method discussed above is a first step in the use of composition-conversion data to probe the importance of charge-transfer complexes in copolymerization processes.[†]

REFERENCES

1. H. Mark, B. Immergut, E. H. Immergut, L. J. Young, and K. I. Beynon, "Copolymerization Reactivity Ratios" in *Polymer Handbook* (J. Brandrup and E. H. Immergut, eds.), Interscience Publishers, New York, 1966.

2. G. E. Ham, *Copolymerization,* Interscience Publishers, New York, 1964.

3. T. Alfrey J., J. J. Bohrer, and H. Mark, *Copolymerization* Vol. VIII of *High Polymers,* Interscience Publishers, New York, 1952.

4. M. Fineman and S. D. Ross, *J. Polymer Sci., 5,* 259 (1950).

5. F. R. Mayo and F. M. Lewis, *J. Amer. Chem. Soc., 66,* 1594 (1944).

6. D. R. Montgomery and C. E. Fry, *J. Polymer Sci.,* Part C, *25,* 59 (1968).

7. D. W. Behnken, *J. Polymer Sci.,* A, *2,* 645 (1964).

8. G. E. P. Box and H. L. Lucas, *Biometrika, 46,* 77 (1959).

9. E. M. L. Beal, *J. Roy. Statist. Soc.,* B22, 41 (1960).

10. P. W. Tidwell and G. A. Mortimer, *J. Polymer Sci.,* A, *3,* 369 (1965).

11. P. W. Tidwell and G. A. Mortimer, *J. Macromol. Sci. Revs. Macromol. Chem.,* C4, 281 (1970).

12. E. Merz, T. Alfrey, and G. Goldfinger, *J. Polymer Sci., 1,* 75 (1946).

13. G. E. Ham, *J. Polymer Sci., 14,* 87 (1954).

14. G. E. Ham, *ibid, 54,* 1 (1961).

15. G. E. Ham, *ibid, 61,* 9 (1962).

16. W. E. Barb, *ibid, 11,* 117 (1953).

17. M. Litt and F. W. Bauer, in *Macromolecular Chemistry, Prague 1965 (J. Polymer Sci. C, 16)* (O. Wichterle and B. Sedlacek, eds.), Interscience, New York, 1967, p. 1551.

[†]The listings and instructions for the three penultimate programs and for the three charge-transfer model programs will be made available upon request.

18. K. Ito and Y. Yamashita, *J. Polymer Sci.*, *A-1*, *4*, 631 (1966).

19. M. Berger and I. Kuntz, *J. Polymer Sci.*, *A*, *2*, 1687 (1964).

20. P. D. Bartlett and K. Nozaki, *J. Amer. Chem. Soc.*, *68*, 1495 (1946).

21. D. Booth, F. S. Dainton, and K. H. Ivin, *Trans. Faraday Soc.*, *55*, 1293 (1959).

22. N. C. Yang and Y. Gaoni, *J. Amer. Chem. Soc.*, *86*, 5022 (1964).

23. G. Henrici-Olivé and S. Olivé, *Z. Phys. Chem. Frankf. Ausg.*, *48*, 35 and 51 (1966).

24. L. P. Ellinger, *Polymer*, *5*, 559 (1964).

25. A. Ledwith, *J. Applied Chem.*, (London), *17*, 344 (1967).

26. S. Iwatsuki and Y. Yamashita, *Makromolek. Chem.*, *89*, 205 (1965).

27. C. Walling, E. R. Briggs, K. B. Wolfstern, and F. R. Mayo, *J. Amer. Chem. Soc.*, *70*, 1537 (1948).

28. A. D. Ketley, *Stereochemistry of Macromolecules*, Vol. 3, Marcel Dekker, New York, 1968, Chapters 1,2.

29. K. Bacon, *New Methods of Polymer Characterization*, Interscience Publishers, New York, 1964.

30. D. W. Marquardt, *Chem. Eng. Progr.*, *55*, 65 (1959).

31. D. W. Marquardt, *J. Soc. Ind. Appl. Math.*, *11*, 431 (1963).

32. R. Foster, *Organic Charge-Transfer Complexes*, Academic Press, New York, 1969.

33. R. S. Mulliken and W. B. Person, *J. Amer. Chem. Soc.*, *91*, 3409 (1969).

34. J. Seyden-Penne, T. Strzalko, and M. Plat, *Tetrahedron Lett.*, 4597 (1965).

35. J. Seiner and M. Litt, *Macromolecules*, *4*, 308, 312, 314, 316 (1971).

36. R. S. Mulliken, *Recl. Trav. Chim. Pays-Bas*, *68*, 147 (1949).

37. L. E. Orgel and R. S. Mulliken, *J. Amer. Chem. Soc.*, *79*, 4839 (1957).

Chapter 6

FORTRAN IV PROGRAMS FOR Q-e SCHEME
CALCULATIONS AND MAPPING

Thane D'Arcy Rounsefell
Charles U. Pittman, Jr.

Department of Chemistry
University of Alabama
University, Alabama

I. INTRODUCTION

In the radical addition copolymerization of vinyl monomers, a
well-known method of classifying the reactivity of a given monomer
is by the semiquantitative Q-e scheme [1,2]. This scheme is used
in polymer chemistry much like the Hammett $\sigma\rho$ equation is used in
organic chemistry to predict rates and equilibria based on substi-
tuent effects [3]. In the Q-e scheme, each monomer has a characteristic value of Q, which is related to resonance stabilization,
and of e, which is related to the polar character of the vinyl
group [4,5]. When a radical addition copolymerization of M_1 and M_2

proceeds via the terminal polymerization model [4], the Q-e scheme
may often be employed to predict, approximately, the reactivity
ratios, r_1 and r_2, for that copolymerization. This has become a
standard method for predicting copolymerization behavior [6].

Equations (1) and (2) relate the values of Q_1, e_1, Q_2, and e_2
to the reactivity ratios $r_1 = k_{11}/k_{12}$ and $r_2 = k_{22}/k_{21}$ of the copolymer Eq. (3). By performing a series of copolymerizations, preferably at optimum experimental conditions [7,8], and fitting that
data to the copolymer equation [4,7-10], in either its integral or
differential forms, the "best" values of r_1 and r_2 are obtained
along with estimates of the upper and lower limits.

$$r_1 = \frac{Q_1}{Q_2} \exp[-e_1(e_1 - e_2)] \tag{1}$$

$$r_2 = \frac{Q_2}{Q_1} \exp[-e_2(e_2 - e_1)] \tag{2}$$

$$\frac{dm_1}{dm_2} = \frac{M_1(r_1M_1 + M_2)}{M_2(r_2M_2 + M_1)} \tag{3}$$

Solving Eqs. (1) and (2) simultaneously, the values of Q_1 and e_1
can be then obtained provided Q_2 and e_2 are known. While this, at
first, appears trivial, it should be noted that the validity and
limitations of the Q-e scheme are many, and these limitations have
been documented [11]. Furthermore, finding r_1 and r_2 is not a
trivial problem [7,8].

Thus, many copolymerization experiments of M_1 with M_2 must be
performed just to evaluate Q_1 and e_1 in copolymerizations with M_2.
Furthermore, the best procedure is to copolymerize M_1 with several
other comonomers (i.e., M_3, M_4, ..., etc.), which cover a range of
Q and e values, and calculate Q_1 and e_1 for each of these processes.
Then, the Q_1 and e_1 values, calculated from each monomer pair (M_1 +
M_2, M_1 + M_3, M_1 + M_4, ..., etc.), are compared to see (1) does the
Q-e scheme approximately hold, (2) what are the "best" values of
Q_1 and e_1, and (3) what is the range of uncertainty which must be

considered. Since r_1 and r_2 values have an uncertainty range, each individual Q-e determination is represented on a Q-e plot (of M_1 and M_2) as an area (Fig. 1). Note that only one of the two areas of intersection has physical significance. One then must compare that intersection area to the other intersection areas which result from copolymerizations of M_1 with M_3, M_4, ..., etc.

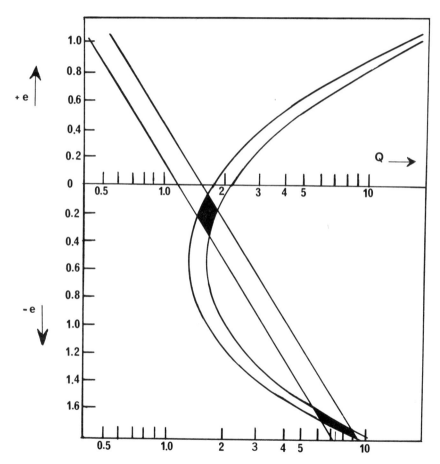

FIG. 1. Q-e plot for the copolymerization of styrene with 2,5-dichlorostyrene.

By making a single Q-e map from each of these copolymer process (i.e., on the same origin), one can rapidly examine all the data and see (1) if the scheme holds and (2) what are the best Q_1 and e_1 values. Figure 2 is just such a map where M_1 is p-chlorostyrene and M_2 is either methyl methacrylate or styrene [4]. Q-e mapping has been described thoroughly elsewhere [11].

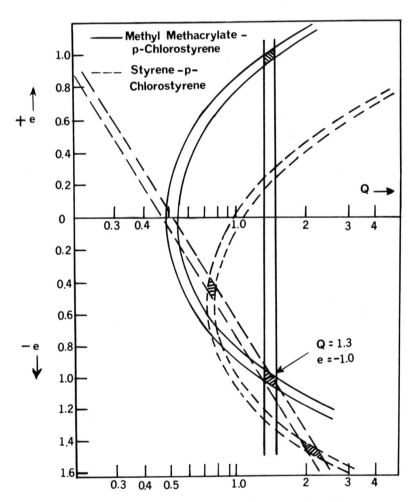

FIG. 2. Q-e map for the copolymerization of p-chlorostyrene with methyl methacrylate and styrene.

The large number of calculations needed in constructing a single Q-e plot, let alone a Q-e map, would be tedious to carry out without computerization. While the programming is routine, such that anyone familiar with computers could write his own, we have provided a FORTRAN IV program in this chapter because of its very broad utility. In this program (QE*PROGRAM'MAIN, Appendix A) Eqs. (1) and (2) are rearranged to (4) and (5). These are then solved simultaneously.

$$Q_1 = \frac{r_1 Q_2}{\exp[-e_1(e_1 - e_2)]} \tag{4}$$

$$Q_1 = \frac{Q_2}{r_2} \exp[-e_2(e_2 - e_1)] \tag{5}$$

Three subroutines have also been provided (see Appendix B). In one, (QE*PROGRAM'QE), Q_1 and e_1 are calculated from input data consisting of Q_2, e_2, r_1, and r_2. In this subroutine, from one to three values of r_1 and r_2 can be entered. That is, the upper and lower error limits of r_1 and r_2, as well as the best r_1 and r_2, can be entered to provide the corresponding Q_1 and e_1 values. Where more than one value of r_1 or r_2 is given, all possible combinations are used in making the calculations.

A second subroutine (QE*PROGRAM'R1R2) permits the calculation of r_1 and r_2 from input consisting of Q_2, e_2, and from one to three values each of Q_1 and e_1. This permits the calculation of the probable range of r_1 and r_2 based on the upper and lower range of Q_1 and e_1 as well as a "best" value. Thus, the effect of uncertainty in the values Q_1 and e_1 on r_1 and r_2 can readily be evaluated.

The final subroutine (QE*PROGRAM'PLOT) both calculates and plots ln Q_1 as a function of e_1 for values of e_1 between 2.5 and -2.5 using 0.05 increments in e_1. The input data includes Q_2, e_2, and both the upper and lower values of r_1 and r_2 may be used. Thus, using upper and lower limits of r_1 and r_2, one obtains two sets of lines which intersect in two places giving areas which define the

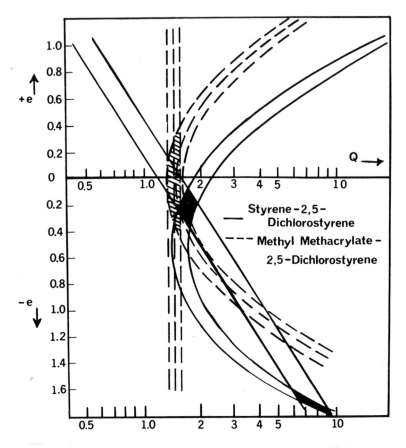

FIG. 3. Q-e map for the copolymerization of 2,5-dichlorosty-
rene with styrene and methyl methacrylate.

range of Q_1 and e_1. Of course, only one of these two solutions has physical significance.

One valuable feature of the PLOT subroutine is that the tabulated $\ln Q_1$ and Q_1 versus e_1 data can be stored. Later, $\ln Q_1$ versus e_1 can be plotted on top of another Q-e plot. This mapping procedure has been arranged so that reactivity-ratio data for the copolymerizations of M_1, with up to 15 other comonomers (i.e., M_2, M_3, ..., M_{15}), may be mapped on the same graph. This is a valuable feature, because the regions of intersection can be compared from one monomer pair to another. Furthermore, the angle of intersection of the plots for an individual monomer pair can be quickly examined. It is quite important that this angle not be too oblique for this can result in a large region of uncertainty in Q and e [13]. Just by looking at r_1 and r_2 data, one cannot necessarily foresee such oblique intersections. Scale expansion is also possible.

A final word should be said concerning r_1, r_2, and Q-e determinations. Anyone designing such a study should pay particular attention to the papers of Tidwell and Mortimer [7,8]. The determination of r_1 and r_2 should be performed recognizing the following criteria:

1. Optimum experimental design is essential (i.e., use of optimum M_1^0/M_2^0 ratios after initially determining r_1 and r_2 crudely).
2. Nonlinear least-squares fitting of composition-conversion data is extremely important.
3. The mathematical technique used to handle the copolymer equation, in its differential or in its integral form, must provide a quantitative statement of the confidence limits (or region) of that r_1, r_2 determination. Unfortunately, most of the data given in such widely used sources as the "Polymer Handbook" [6] has not been obtained using these criteria. This may account for a portion of the past failures of the Q-e scheme.

APPENDIX A

Qe* Program•Main

```
C ********* Q-E PROGRAM *********
C FOR INSTRUCTIONS RUN PROGRAM WITH ONE BLANK DATA CARD.
C PROGRAMED BY THANE ROUNSEFELL      UNIVERSITY OF ALABAMA     12/73
C      VER. 1-3   12/73
       COMMON X
6 READ(5,5000,END=9999) X
5000 FORMAT(F10.6)
     IF(X .LE. 0.0) GO TO 5
1 IPICK=IFIX(X+.1)
     IF(IPICK .LT. 1 .OR. IPICK .GT. 4) GO TO 7
     GO TO (2,3,4,5),IPICK
2 CALL PLOT
     GO TO 1
3 CALL QE
     GO TO 1
4 CALL R1R2
     GO TO 1
5 WRITE(6,5001)
     WRITE(6,5002)
     WRITE(6,5003)
     WRITE(6,5004)
     WRITE(6,5005)
     GO TO 6
7 WRITE(6,5012) X
5012 FORMAT('1','ERROR -- THE FOLLOWING NUMBER WAS USED AS A CALCULATIO
1N SELECTOR ',F10.6,/,' THE CALCULATION SELECTOR MUST BE',/,' 1.0 P
1LOT',/,' 2.0 Q-E',/,' 3.0 R1R2',/,' 4.0 INSTRUCTIONS')
```

```
9999 STOP
5001 FORMAT('1',21X,   'Q-E PROGRAM -- INSTRUCTIONS',//,6X,
    1 'THIS PROGRAM CONSITS OF THREE CALCULATIONS.   THE FIRST CALCULATI
    20N',/,1X,       'IS A Q-E PLOT, THE SECOND IS A CALCULATION OF Q1'
    3 ,' AND E1 FROM R1, R2,',/,1X,'Q2, AND E2, AND THE THIRD IS A',
    4 ' CALCULATION OF R1 AND R2 FROM Q1, E1,',/,1X,  'Q2 AND E2.',//
    5, 6X,'TO USE THIS PROGRAM IT IS NECESSARY TO USE ONE DATA CARD TO
    6 SELECT',/,1X,'THE PROPER CALCULATION, FOLLOWED BY THE DATA ',
    6'FOR THE CALCULATION. THE',/,1X,                      'CARD WH
    7ICH SELECTS THE CALCULATION MUST HAVE THE VALUE 1., 2., 3., OR'
    8 ,/,1X,  '4. PUNCHED IN THE FIRST 10 COLUMNS.  THE CALCULATION ',
    9 'SELECTORS HAVE',/,1X,  'THE FOLLOWING MEANINGS.',//,23X,
    1 '1.0 -- Q-E PLOT' ,/,23X,  '2.0 -- Q-E CALCULATION',/,23X,
    1 '3.0 -- R1-R2 CALCULATION' ,/,23X, '4.0 -- PRINTS INSTRUCTIONS'
    2 ,//,6X,   'NOTE THAT ALL INPUTS (EXCEPT LABELS) ARE REAL NUMBERS.'
    2 ,' ALL INPUT' ,/,1X, 'VALUES TAKE 10 CARD COLUMNS EACH EXCEPT ',
    3 'LABELS WHICH TAKE ONE FULL' ,/,1X, 'CARD.  REMEMBER THAT Q1, E1
    4, AND R1 ARE ASSOCIATED WITH MONOMER 1 AND' ,/,1X, 'Q2, E2 AND R2'
    5 ' ARE ASSOCIATED WITH MONOMER 2')
5002 FORMAT('0',//,27X,'INPUT FOR Q-E PLOT',//,1X,'CARD A -- LOWER R1,
    1UPPER R1, LOWER R2, UPPER R2, Q2, E2, Q1X', E1X'/,1X,'CARD B --'
    2 ' LABEL' ,/,1X,'CARD C -- BLANK' ,/,1X,  'CARD D -- NEGATIVE VAL
    6UE IN COLUMNS 1-10',//,1X,  'CARD B MUST ALWAYS FOLLOW CARD A.'
    4 ,/,1X,   'CARDS C AND D ARE ONLY USED WHEN NEEDED.',/,6X,  'A PLOT
    5 IS PRINTED FOR EACH CARD A AND THIS PLOT IS ALSO SAVED.' ,/,1X,
    6 'CARD C CAUSES ALL SAVED PLOTS TO BE SUPERIMPOSED IN ONE CUMULATI
    7VE' ,/,1X, 'PLOT.  CARD D CAUSES ALL SAVED PLOTS TO BE ERASED. 15
    8 PLOTS IS THE' ,/,1X,'LIMIT THAT CAN BE SAVED.',/,1X,
    9'Q1X AND E1X (OPTIONAL) MAY BE USED TO CHANGE THE PLOT DIMENSIONS'
```

```
      1,' FROM' ,/,1X, 'THE STANDARD.  Q1X IS THE LOWER LIMIT OF Q1. '
      2,' 'E1X IS THE MAXIUM OF E1.',/,1X,'WHEN Q1X IS GIVEN THE RAN
      3GE OF Q1 ON THE PLOT IS EXPANDED TO 15')
 5003 FORMAT('0',//,23X,'INPUT FOR Q-E CALCULATIONS' ,//,1X, 'CARD E --
      1 R1, LOWER R1, UPPER R1, R2, LOWER R2, UPPER R2, Q2, E2' ,/,1X,
      2 'CARD F -- LABEL' ,//,1X, 'THE LOWER AND UPPER VALUES OF R1 AND '
      3 'R2 ARE OPTIONAL.  IF THEY ARE' ,/,1X, 'GIVEN ALL POSSIBLE '
      4 'COMBINATIONS OF R1 AND R2 WILL BE USED TO CALCULATE' ,/,1X, 'Q1
      5AND E1.  EITHER 1, 2, OR 3 VALUES OF R1(R2) CAN BE ENTERED.  IF 2'
      6 ,/,1X, 'VALUES FOR R1(R2) ARE ENTERED THE UPPER R1(R2) MUST BE'
      7 ,' THE BLANK VALUE' ,/,1X, 'CARD F MUST FOLLOW CARD E')
 5004 FORMAT('1',22X,'INPUT FOR R1-R2 CALCULATION',//,1X, 'CARD G -- '
      1 'Q1, Q1, E1, E1, Q2, E2' ,/,1X, 'CARD H -- LABEL' ,//,
      2 6X, 'THE SECOND AND THIRD VALUES OF Q1(E1) ARE OPTIONAL.  IF '
      3 'THEY ARE' ,/,1X, 'GIVEN ALL POSSIBLE COMBINATIONS OF Q1 AND E1'
      4 ,' WILL BE USED TO CALCULATE' ,/,1X, 'R1 AND R2.' ,//,1X,
      5 ,' CARD H MUST FOLLOW CARD G.')
 5005 FORMAT('0',//,30X, 'SAMPLE INPUT' ,///,3X, '1.0  -- SELECTS Q-E
      1PLOT' ,/,2(1X,'CARD A',/,1X,'CARD B -- PRINTS ONE PLOT',/) ,1X,
      2 'CARD C -- PRINTS CUMULATIVE PLOT OF TWO PREVIOUS PLOTS' ,/,1X
      2 , 'CARD A' ,/,1X, 'CARD B -- PRINTS ONE PLOT' ,/,1X
      3 'CARD C -- PRINTS CUMULATIVE PLOT OF THREE PREVIOUS PLOTS'
      4 ,/,1X, '4.0  -- PRINTS INSTRUCTIONS',/,1X,' 3. -- ',
      5 'SELECTS R1-R2 CALCULATION' ),/,1X, 'CARD G' ,/,1X, 'CARD H --
      6 R1-R2 CALCULATION' ),/,1X,' 1. -- SELECTS Q-E PLOT' ,/,1X
      7 ,' CARD A' ,/,1X, 'CARD B -- PRINTS ONE PLOT' ,/,1X, 'CARD'
      9 ,' D  -- ERASES CUMULATED PLOTS' ,/,1X, ' 2.0  -- Q-E' SELECTS
      1 Q-E CALCULATION' ,/,1X, 2('CARD E' ,/,1X, 'CARD F  -- Q-E'
      1 ,' CALCULATION' ,/,1X))
      END
```

APPENDIX B

Qe* PROGRAM·Qe

```
      SUBROUTINE QE
      REAL X(3),Y(3),ALABEL(20)
      EQUIVALENCE (X(1),R1),(Y(1),R2)
      COMMON R1
    1 READ(5,100,END=99) X,Y,Q2,E2
  100 FORMAT(8F10.6)
      IF(Q2 .LE. 0.0) RETURN
      READ(5,104,END=99) ALABEL
  104 FORMAT(20A4)
      WRITE(6,105)
      WRITE(6,101) ALABEL,Q2,E2,X,Y
      II=1
      JJ=1
      IF(X(2) .GT. 0.0) II=2
      IF(X(3) .GT. 0.0) II=3
      IF(Y(2) .GT. 0.0) JJ=2
      IF(Y(3) .GT. 0.0) JJ=3
    5 DO 10 I=1,II
      DO 10 J=1,JJ
      A=X(I)*Y(J)
      IF(A .LE. 0. .OR.A.GE. 1.0) GO TO 20
      AL=SQRT(-ALOG(A))
      EX=E2+AL
      EY=E2-AL
      QX=(Q2/Y(J))*EXP(-E2*(E2-EX))
      QY=(Q2/Y(J))*EXP(-E2*(E2-EY))
      WRITE(6,102) X(I),Y(J),A,QX,EX,QY,EY
      GO TO 10
```

```
   20 WRITE(6,103) X(I),Y(J),A
  101 FORMAT('0',20A4,//,1X,    'Q2 =',F10.6,8X,'E2 =',F10.6,//,1X,'R1 ='
     1 ,F10.6,5X,F10.6,5X,F10.6,//,1X,   'R2 =',F10.6,5X,F10.6,5X,F10.6,
     2 ///,8X,'R1',10X,'R2',  8X,'R1 X R2',7X,2( 'Q1' ,10X,'E1',10X))
  102 FORMAT('0',7F12.6)
  103 FORMAT('0',3F12.6,10X,'CAN NOT CALCULATE SQRT(-LOG10(R1 X R2))')
  105 FORMAT('1',15X,   'Q-E PROGRAM -- CALCULATION OF Q1 AND E1 FROM'
     ,         'R1, R2, Q2 AND E2')
   10 CONTINUE
      GO TO 1
   99 STOP
      END
```

Qe* Program·R1R2

```
      SUBROUTINE R1R2
      INTEGER ALABEL (20)
      REAL X(3),Y(3)
      COMMON X
    1 READ(5,100,END=99) X,Y,Q2,E2
  100 FORMAT(8F10.6)
      IF(Q2 .LE. 0.0) RETURN
      READ(5,103,END=99) ALABEL
  103 FORMAT(20A4)
      WRITE(6,132)
      WRITE(6,131) ALABEL,Q2,E2,X,Y
      L=1
      M=1
      IF(X(2) .GT. 0.0) L=2
      IF(X(3) .GT. 0.0) L=3
```

```
      IF(Y(2) .GT. 0.0) M=2
      IF(Y(3) .GT. 0.0) M=3
    5 DO 10 I=1,L
      DO 10 J=1,M
      R1=(X(I)/Q2)*EXP(-Y(J)*(Y(J)-E2))
      R2=(Q2/X(I))*EXP(-E2*(E2-Y(J)))
      WRITE(6,102) X(I),Y(J),R1,R2
  102 FORMAT('0',28X,4F15.6)
  131 FORMAT('0',20A4,//,1X,    'Q2 =',F10.6,8X,   'E2 =',F10.6,//,1X,
     1  'Q1 =', F10.6,5X,F10.6,5X,F10.6,//,1X,   'E1 =',F10.6,5X,F10.6,
     3 5X,F10.6,//,38X,   'Q1',13X,   'E1',13X,   'R1',13X,   'R2')
  132 FORMAT('1',15X,   'Q-E PROGRAM -- CALCULATION OF R1 AND R2 FROM Q1,
     1 Q2, E1 AND E2')
   10 CONTINUE
      GO TO 1
   99 STOP
      END


                    Qe* Program•Plot

      SUBROUTINE PLOT
      REAL R(4),Q1(4),LQ1(4)   ,EX(101)
      INTEGER*2 K(101,60)  ,MARK(115),STAR/1H*/
      INTEGER*2   L(115)/115*1H /,DOT/1H./,BLANK/1H /
      INTEGER*4 ALABEL(20)
      COMMON R
      DO 90 I=1,101
   90 EX(I)=(1.-I)/20.+2.5
      DO 91 I=1,115
   91 MARK(I)=BLANK
      MARK(86)=STAR
```

```
      MARK(95)=STAR
      MARK(100)=STAR
      MARK(104)=STAR
      MARK(106)=STAR
      END1=.001
      END2=10.
      XLQ1=6.9
      RANGE=9.2
    5 JX=0
    4 READ(5,100,END=2) R,Q2,E2,Q1X,E1X
  100 FORMAT(8F10.6)
      IF(R(1))5,3,23
   23 IF(Q2 .LE. 0.0) RETURN
      IF(E1X)25,26,25
   26 IF(Q1X .GT. 0.0) GO TO 27
      GO TO 24
   25 E1X=E1X-2.5
      DO 29 I=1,101
   29 EX(I)=EX(I)+E1X
      JX=0
      GO TO 26
   27 JX=0
      XLQ1=-ALOG(Q1X)
      RANGE=XLQ1+ALOG(Q1X+15.)
      DO 32 I=1,115
   32 MARK(I)=BLANK
      END1=Q1X
      END2=Q1X+15.
      DO 31 I=1,50
      KTEMP=((ALOG(FLOAT(I))+XLQ1)/RANGE)*115. + .5
      IF(KTEMP .LT. 1) GO TO 31
```

```
      IF(KTEMP .GT. 115) GO TO 24
      MARK(KTEMP)= STAR
   31 CONTINUE
   24 READ(5,107) ALABEL
  107 FORMAT(20A4)
      WRITE(6,105)
      WRITE(6,101)         ALABEL,Q2,E2,R
  101 FORMAT('0',20A4,//,          '   Q2 =',F10.6,'   E2 =',F10.6,
     1//,22X,'LOWER R1',16X,'UPPER R1',16X,'LOWER R2',16X,'UPPER R2',/,
     221X,F10.6,3(14X,F10.6),//,8X,'E1',   4(9X,'Q1',8X,'LN Q1'),/)
      IF(JX .GT. 56) JX=0
      DO 20 I=1,101
      E1=EX(I)
      EXPX=EXP(-E1*(E1-E2))
      EXPY=EXP(-E2*(E2-E1))
      DO 10 J=1,2
      JZ=J+2
      Q1(J)=R(J)*Q2/EXPX
   10 Q1(JZ)=(Q2/R(JZ))*EXPY
      DO 15 J=1,4
      IF(Q1(J) .LE. 0.0) GO TO 8
      LQ1(J)=ALOG(Q1(J))
      KTEMP=((LQ1(J) + XLQ1/RANGE)*115. + .5
      IF(KTEMP .LT. 1 .OR. KTEMP .GT. 115) GO TO 8
      K(I,J+JX)=KTEMP
      GO TO 15
    8 K(I,J+JX)=1
   15 CONTINUE
      WRITE(6,102) E1,Q1(1),LQ1(1), Q1(2),LQ1(2),Q1(3),LQ1(3),Q1(4),LQ1(
     14)
C ***    ***    ***    ***    ***    ***    ***    ***
C THE NEXT 2 STATEMENTS ARE FOR PRINTER CONTROL
      IF(I .EQ. 46) WRITE (6,109)
```

```
  109 FORMAT('1')
C ***     ***     ***     ***     ***     ***     ***     ***
  102 FORMAT(' ',9F12.6)
   20 CONTINUE
      JXX=JX+1
      JXY=JX+4
C ***     ***     ***     ***     ***     ***     ***     ***
C THE FOUR CALLS TO PRTCN AND PRTCN2 ARE FOR PRINTER CONTROL.  THE PLOT
C LOOKS BETTER WITH PRINTER CONTROLS BUT THEY MAY BE REMOVED WITHOUT
C AFFECTING CALCULATIONS.  (SEE COMMENTS IN SUBROUTINE PRTCN)
C ***     ***     ***     ***     ***     ***     ***     ***
      CALL PRTCN
      WRITE(6,105)
C THE NEXT THREE STATEMENTS SHOULD BE REPLACED WITH THE PRECCEDING
C STATEMENT IF NO PRINTER CONTROL IS USED.
      WRITE(6,110)
  110 FORMAT('1',/////,15X,'Q-E PROGRAM -- CALCULATION OF Q-E PLOT FROM
     1 R1, R2, Q2 AND E2')
      WRITE(6,106) ALABEL
      WRITE(6,103) END1,END2,MARK
  103 FORMAT('0',3X,F6.3,41X,'LN Q1'   ,T117,F6.3,/,1X,' E1  ',115A1)
      DO 30 I=1,101
      DO 40 J=JXX,JXY
      II=K(I,J)
   40 L(II)=DOT
      WRITE(6,104) L  ,EX(II)
  104 FORMAT(' ',5X,115A1,T3,F5.2)
      DO 50 J=JXX,JXY
      II=K(I,J)
```

```
 50 L(II)=BLANK
 30 CONTINUE
    CALL PRTCN2
    JX=JX+4
    GO TO 4
  3 WRITE(6,105)
    WRITE(6,108)
    WRITE(6,103) END1,END2,MARK
    CALL PRTCN
    DO 80 I=1,101
    DO 70 J=1,JX
    II=K(I,J)
 70 L(II)=DOT
    WRITE(6,104) L ,EX(I)
    DO 75 J=1,JX
    II=K(I,J)
 75 L(II)=BLANK
 80 CONTINUE
    CALL PRTCN2
    GO TO 4
  2 STOP
105 FORMAT('1',15X, 'Q-E PROGRAM -- CALCULATION OF Q-E PLOT FROM',
   1 ' R1, R2, Q2 AND E2')
106 FORMAT('0',20A4)
108 FORMAT('0','CUMULATIVE PLOT')
    END
```

```
.  THIS ASSEMBLER SUBROUTINE CONTROLS THE SPACING ON THE PRINTER
.  TO MAKE THE Q-E PLOT RUN TOGETHER.  IT IS FOR UNIVAC COMPUTERS
.  ONLY.  IT MAY BE REMOVED OR REWRITTEN FOR OTHER MACHINES.
.  (SEE PLOT SUBROUTINE)
.  CALL PRTCN    CHANGES PRINTER TO PRINT 66 LINES PER PAGE
.  CALL PRTCN2   CHANGES PRINTER BACK TO PRINT 57 LINES PER PAGE
$(1)
PRTCN*    AXR$
          L        A0,(02,BUFF)
          ER       PRTCN$
          J        1,X11
PRTCN2*   L        A0,(02,BUFF2)
          ER       PRTCN$
          J        1,X11
$(0),BUFF          'M,66,0,0. .  '
BUFF2              'M,66,6,3. .  '
          END
```

REFERENCES

1. T. Alfrey and C. C. Price, *J. Polymer Sci., 2*, 101 (1947).

2. C. C. Price, *ibid, 3*, 772 (1948).

3. L. P. Hammett, "Physical Organic Chemistry," Chap. VII, McGraw-Hill, New York.

4. T. Alfrey, J. J. Bohrer, and H. Mark, "Copolymerization," Vol. VIII of *High Polymers,* Chaps. II-IV, Interscience Publishers, Inc., New York, 1952.

5. T. Tsuruta and K. F. O'Driscoll (eds.), "Structure and Mechanism in Vinyl Polymerization," Chaps. 1 and 2, Marcel Dekker, Inc., New York, 1969.

6. J. Brandrup and E. H. Immergut (eds.), "Polymer Handbook," Interscience Publishers, New York, 1966, Sections II6-II8.

7. P. W. Tidwell and G. A. Mortimer, *J. Polymer Sci., A, 3*, 369 (1965).

8. P. W. Tidwell and G. A. Mortimer, *J. Macromol. Sci. Revs. Macromol. Chem., C4*, 281 (1970).

9. D. R. Montgomery and C. E. Fry, in "The Computer in Polymer Science," *J. Polymer Sci., C, 25*, J. B. Kinsinger, ed.), Interscience, New York, 1968, p. 59.

10. M. Fineman and S. D. Ross, *J. Polymer Sci., 5*, 259 (1950).

11. See Ref. 4, Chap. IV, part 4.

Chapter 7

A RHEOMETER FOR CHARACTERIZATION OF NONLINEAR
AND TIME-DEPENDENT FLUIDS

Tyan-faung Niu[*]

Irvin M. Krieger

Department of Macromolecular Science
Case Western Reserve University
Cleveland, Ohio

---

[*]Current affiliation: Photo and Repro Division, GAF Corporation,
Binghamton, New York.

I. INTRODUCTION

Polymer solutions, melts, and colloidal dispersions are very different in their rheological behavior from other fluids, as they exhibit non-Newtonian effects in steady-state shearing flow, time-dependent behavior, normal stress, viscoelastic behavior, and other unusual effects. These phenomena complicate the design of polymer processing equipment. To handle such an engineering problem, information on their mechanical properties is needed. Therefore it is one of the major goals of the polymer rheologist to find a good technique to characterize the flow behavior of such fluids, by measuring the relation between the applied shear strain and the response shear stress over a wide range of time and temperature. The rheometer system presented here has been built around a PDP-8/L computer to provide a versatile and efficient method for the characterization of nonlinear and time-dependent fluids.

When a linear fluid is subjected to a shear strain varying sinusoidally with time, the response stress is also sinusoidal, whereas a nonlinear fluid under the same shear strain will respond with a nonsinusoidal periodic stress at the same frequency. By means of Fourier analysis one can resolve each nonsinusoidal variation into a sum of a "fundamental" harmonic variation and harmonic "overtones." These Fourier coefficients can be used to characterize the viscoelastic properties of nonlinear fluids. Due to the complications of nonlinear calculations, small amplitude oscillatory experiments are widely used to characterize the flow properties of viscoelastic fluids. The amplitude of the input oscillatory motion is usually kept sufficiently small that the theory of linear viscoelasticity is valid, i.e., the stress-strain amplitude ratio, though a function of frequency, is independent of the strain amplitude.

When a reduction in magnitude of viscosity of a fluid occurs reversibly and isothermally with a distinct time-dependency on application of shear strain, the phenomenon is called thixotropy. Many solid-liquid suspensions, such as greases and clay-water systems [1,2], exhibit marked time dependency in their shear stress-shear

rate relations. The "hysteresis" type of experiment has been widely
used in the investigation of the rheological properties of such kinds
of materials [3]. It is based on the fact that at high shear rates
the structure is more readily broken down, whereas at low shear rates
the structure builds up. The test substance, which is typically contained in the annulus between two coaxial cylinders, is subjected to
a programed shear in which the drive cylinder is accelerated from
rest to a predetermined rotational speed, and then decelerated to
rest again. The loop formed by plotting torque against rotational
speed for a complete cycle is considered a measure for the thixotropic behavior of the material. Although some complication was
pointed out in the interpretation of the hysteresis loop in terms of
thixotropic properties [2,4], this technique is still considered
useful in industry as a rapid means for ranking different thixotropic
fluids and for quality control.

    Dodge and Krieger [5,6] have investigated the use of oscilla-
tory shear for the study of time-dependent fluids. The technique is
based on the fact that, if one varies the shear rate sinusoidally at
constant amplitude, then at low frequencies one will always retrace
the equilibrium curve as obtained by steady-state shear experiments.
As one increases the frequency to the order of the reciprocal of the
largest breakdown or buildup time, one deviates from the equilibrium
curve.

    By solving the equation of motion, a prediction in the form of
Fourier coefficients can be made of the oscillatory behavior of a
viscometer, based on steady-state experimental data of viscosity as
a function of shear rate only. Then, at low frequencies the predicted and observed harmonic amplitude and phase angles of the
stress should be in agreement. At frequencies where the time dependency of the viscosity is important, there will be a discrepancy between the observed and the predicted results. Their experimental
data were obtained using the Weissenberg rheogoniometer [7], while
finite difference techniques predicted the oscillatory data for nonlinear, purely viscous materials. The experimental and predicted

data are in good agreement for purely viscous fluids. Large discrepancies were shown by thixotropic and viscoelastic fluids, which
indicate time dependency.

Although their results indicate the suitability of oscillatory
testing procedures for investigation of time-dependent materials,
Dodge and Krieger encountered difficulties in analyzing and inter-
preting some of their data, due to the limitations of their apparatus. They pointed to the need for: (1) continuously variable
driving frequencies, [2] larger shear amplitudes, particularly at
low frequencies, (3) upward extension of the frequency range, and
(4) better methods of data acquisition and analysis, so that signal
averaging could be conveniently employed. The use of a servomotor
drive system and on-line computer were suggested.

To accomplish these objectives, a rotational rheometer was
designed to operate in conjunction with an on-line minicomputer.
Computer programs in assembly language have been written for both
oscillatory-shear and hysteresis-loop experiments. In the oscillatory shear mode, the program automatically scans a range of driving frequencies (the limits are from 9×10^{-3} Hz to 156 Hz) with
shear strain amplitude or strain rate amplitude kept constant. A
steady shear rate can be superposed on the oscillatory shear. The
stress and strain-rate data are Fourier-analyzed for the first five
harmonics. The program can also compute the real and imaginary
parts of the complex shear modulus, G' and G'', based on the first
harmonic, and graph these moduli as functions of frequency on an
X-Y plotter. In the hysteresis-loop mode, it scans a range of
shear rates from zero to a selected maximum and then back to zero,
and then draws the hysteresis loop based on the computed shear
stress-shear rate data. At a sufficiently slow scanning rate, the
hysteresis-loop program can be used to characterize steady-state
flow behavior.

This chapter describes the instrumentation, the generation of
forcing function signals, the data acquistion, and the analysis of
the response signals. Results of preliminary tests of each mode

of operation are presented. It also discusses the results from the time-dependent study on latex suspensions. The somewhat lengthy computer programs have been listed elsewhere [15] and will not be presented here.

II. DESCRIPTION OF THE SYSTEM

A. Apparatus

A diagram of the Couette type rheometer which is to be attached to the signal processing instrument is shown in Fig. 1. The liquid is contained in the annular gap formed by the coaxial stainless steel cylinders. The partially hollow inner cylinder is 6.509 cm long and of outer diameter 6.350 cm. The outer cylinder is 8.414 cm long and 6.429 cm diameter (both inside dimensions), with a slight shoulder at the top to permit a small amount of overfilling. The volume of liquid required is approximately 60 cm^3. By driving the outer cylinder instead of the inner cylinder, Taylor vortex instabilities [16] are avoided, but a special seal is needed where the drive shaft enters the bottom of the water bath. An adapter was machined to connect the outer cylinder to the shaft of a servomotor. The adaptor carries at its midpoint an inverted cup, which is immersed in the mercury contained between two short coaxial cylinders at the bottom of a constant temperature bath to form a mercury seal.

The gearless servomotor (Photocircuits Corporation, Glen Cove, New York, model U12M4T) is capable of speed ranges from 5000 rpm down to zero. With a coupled tachometer and a differential servo amplifier, the motor speed can be made accurately proportional to the voltage of a control signal (5.443 VDC/1000 rpm). Low-inertia "printed-circuit" rotors, formed by stamping the metal winding onto the surfaces of a fiberglass disk, make it possible for the motor speeds to follow rapid changes in the control voltage. The motor was tested in our laboratory and gave sinusoidal oscillation up to about 1 kHz when controlled by an audio signal generator.

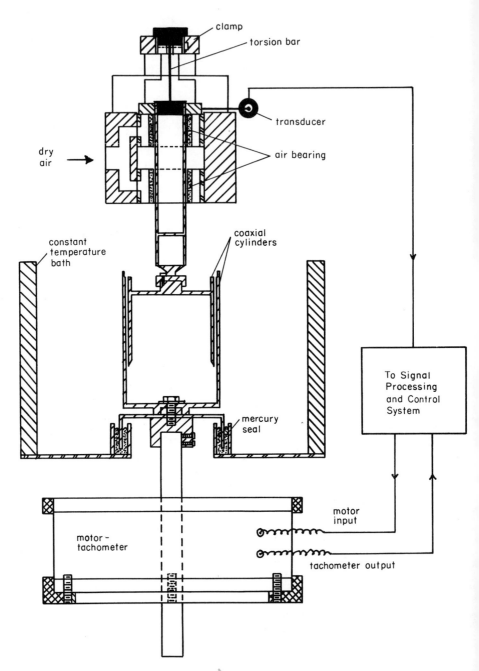

FIG. 1.   Schematic diagram of the rheometer.

The motion of the inner cylinder, caused by the stress transmitted through the fluid from the outer cylinder, is indicated by
the output voltage signal of a variable inductance displacement
transducer to which a transducer meter (Boulton Paul Aircraft Ltd.,
Wolverhampton, England. Type EP 597) is connected. The inner
cylinder is supported by a hollow rotor which in turn is supported
by one of several interchangeable torsion bars. The rotor and the
torsion bars were purchased as parts of the Weissenberg rheogoniometer (Sangamo Weston Ltd., Bognor Regis, England. Model R 16).
They are described in detail in the manufacturer's manual [7]. The
rotor is centered by two air bearings (Apex Bearing Co., Hudson,
Ohio. Part No. AAB-1628-8) to minimize friction. Both the motor
outer-cylinder assembly and the inner-cylinder air-bearing torsion-
bar assembly were mounted on carriers of a vertical dovetail slide;
a lead screw permits the upper carrier to be raised or lowered.

The output signals from the transducer meter and tachometer
are passed through a low-pass "twin filter unit" to remove noise
during oscillatory measurements. Since the filter unit causes
attenuation and phase shift of the signal, the filters in the twin
filter unit were matched to provide the same attenuation and phase
shift to both the speed signal from the tachometer and the torsion
signal from the transducer.

The constant temperature bath, with its water level covering
the entire outer cylinder to a point above the sample level, provides control accuracy of $\pm$ 0.025°C.

For the air bearing, compressed air from the laboratory air
supply is first fed to a refrigeration type air dryer (Ingersoll-
Rand Co., New York, New York, Model 2) to remove moisture, dirt,
oil vapor and other contaminants. It then passes through a cotton
filter unit mounted at the back of the instrument for further puri-
fication. An air pressure at the bearing of 20 psi was found to
cause minimum friction between rotor and bearing.

The effective length of the inner cylinder was determined using
a standard Newtonian oil to be 6.647 cm; this indicates that the end
effect contributes 2.1% of the measured torque.

Table 1 lists the torsion bar constants, K_θ, and the resonant frequencies for the torsion bars available. The moment of inertia I_o of the inner cylinder assembly was calculated from the measured resonance frequency and the known torsion bar constant, which led to $I_o = 2283$ g·cm^2.

B. Data Processing Instrumentation

The data processing equipment consists of a central processor and peripheral devices to communicate with the apparatus and with the experimenter. Figure 2 is a schematic diagram of the equipment along with the paths of communication among its components. The Central Processor is a PDP-8/L computer (Digital Equipment Corp., Maynard, Massachusetts) which is a 12-bit, 4,096-word, 1.6 microsecond cycle time computer with an input-output bus and an ASR 33 teletype with paper tape unit. The computer's memory unit has been expanded by adding an extra 4,096-word memory module.

Loading a long paper tape into the PDP-8 core memory with the reader of the teletype unit is very time consuming (it inputs at ten characters per second maximum). A high-speed paper tape reader was installed which inputs information photoelectrically at a rate of 300 characters per second.

TABLE 1

Resonant Frequencies and Torsion Constants
with the Torsion Bars

Diameter of torsion bar (inches)	K_θ (dyn-cm/radian)	K_T (dyn-cm/mil[a])	Resonant frequency (cycles/sec)
1/32	4.958×10^5	1.259×10^2	2.352
1/16	8.474×10^6	2.152×10^3	9.667
1/8	1.240×10^8	3.404×10^4	38.57

[a]Transducer deflection is indicated in mils.

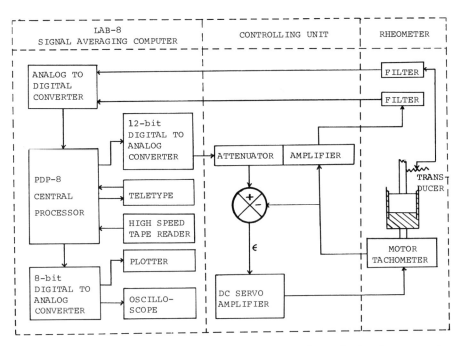

FIG. 2. Schematic diagram showing coupling of the rheometer to LAB-8/L minicomputer.

A special input-output device, the AX08 laboratory peripheral unit, was connected to the central processor; the combination of the PDP-8/L with the AX08 is marketed by Digital Equipment Corp. as the LAB8-L Signal Averaging Computer. It includes an analog-to-digital converter (ADC) which can monitor four analog inputs and convert them to 8-bit signed numbers and an 8-bit digital-to-analog converter (DAC) which converts digital information from the central processor to analog output voltages for the oscilloscope (Tektronix, Inc., Portland, Oregon, Type RM-503) display; or for an X-Y recorder (Hewlett-Packard Co., Moseley Division, Pasadena, California, Model 7035B) to plot cartesian coordinate graphs. It is also equipped with clocks which generate timing pulses to trigger the ADC or DAC conversions. A Polaroid Land camera (Tektronix, Inc., Model C-12) was mounted on the oscilloscope for photographing the displays.

Since the DAC provided in the AX08 unit produces an 8-bit analog output from 0 to -10 V which is not suitable in range or accuracy for the oscillatory drive of the servomotor, a separate 12-bit DAC (Datel Systems, Inc., Canton, Mass., Model DAC-HB 12B) was installed, whose output voltage is bipolar and ranges from -5 to 5 V. It is used for the generation of an analog waveform to drive the servomotor from a function stored in computer's memory. To produce even better precision, the 12-bit DAC is always operated near its full output of ±5V. Whenever small control voltages are needed, the output is reduced by an attenuator.

The motor is driven from a differential servo amplifier (Control Systems Research, Inc., Pittsburgh, Pa., Model 500), which compares the tachometer output voltage with the input control voltage, accelerates the motor with this difference voltage until it reaches a speed such that the "difference" is just sufficient to provide the necessary drive voltage to the motor; the tachometer voltage is then very nearly equal to the input voltage. In steady-state rotation, a speed of 918 rpm will produce a tachometer output of 5 V; hence a 5-volt control voltage is needed to produce a motor speed of 918 rpm. Since the 8-bit ADC accepts voltages ranging from -1 to 1 V, an amplifier-attenuator was built to provide input voltages which can be converted with adequate precision.

For the selection of the desired attenuation rate for the motor control signal and the amplification rate for the input signal to the ADC, there are 9 steps on the attenuation-amplification switch. The attenuation rate ranges from 0.0001 to 1; in each step, the product of attenuation rate and amplification rate is always 0.2. For example, with a 5-volt driving voltage fed into the motor, the tachometer signal is attenuated to 1 V and input to the ADC.

C. The Forcing Function Drive Signal

1. Sine Waves

  a. Constant Frequency Driving. To generate a sinusoidal drive signal by the central processor, a memory block of N words are reserved for data storage (with N chosen as 32, 64, 128, 256, or 512).

The value of the function sin ψ_i is computed for each angle ψ_i = $2\pi i/N$ radians in the first quadrant ($0 \leq \psi_i \leq \pi/2$, where i = 0, 1, 2, ..., N/4; after multiplication by the amplitude factor A, the sines are stored in consecutive locations of the reserved memory block. The amplitude factor A is equal to the desired maximum output voltage from the 12-bit DAC. Without further computation, the sin ψ_i data of the second quadrant are simply stored by using the data in the first quadrant based on the relation sin ψ_i = $\sin(\pi - \psi_i)$. For sin ψ_i values in the second half cycle, since sin ψ_i = $-\sin(\psi_i - \pi)$, the data in the first half are used with negative sign to complete one full cycle of the sine function.

The sine data are called out sequentially on pulses from the computer's crystal clock (which pulses every 100 μsec) at intervals appropriate to the desired frequency, and sent to the 12-bit DAC. The analog voltage simulated a sinusoidal drive signal with N steps. Assuming the number of clock pulses between trigger pulses is M, the period of the sine wave would thus be N x M x 100 μsec. The period can be shortened by selecting fewer steps for the cycle of sine data, or by waiting for fewer clock pulses between steps. The program provides about 20,000 such combinations, giving frequencies ranging from 0.009 Hz to 156 Hz.

Since the peak drive voltage is A, the sinusoidal drive signal is expressed as A sin(2πft), where f and t are the frequency in Hz and time in seconds. As the motor speed is linearly proportional to the drive voltage, the angular speed of the motor during sinusoidal drive is

$$V_\theta = \frac{A}{K_V} \sin(2\pi ft) \tag{1}$$

where the constant K_V is defined as the ratio of motor speed to control voltage. Integrating V_θ over t, the displacement θ (in radians) of motor shaft and outer cylinder is obtained (θ = 0 at t = 0) as

$$\theta = \frac{-A}{2\pi fK_V} \cos(2\pi ft)$$

or

$$\theta = \frac{A}{2\pi f K_V} \sin(2\pi ft - \frac{\pi}{2}) \tag{2}$$

so that the displacement signal of the outer cylinder lags by 90 degrees the driving voltage signal; the amplitude is $\theta_o = A/(2\pi K_V f)$. The shear-rate amplitude can be estimated at the radius of the outer cylinder times the outer cylinder amplitude times the frequency divided by the gap distance. For the experimental conditions used, the expression is $\dot{\gamma}_o = 509\theta_o f$.

b. Frequency Scanning. This rheometer system is programmed to carry out measurements over a frequency range. After completing an oscillatory experiment at one frequency, the computer will automatically start the next oscillatory experiment at a higher frequency multiplied by the typed "frequency increasing factor." It achieves such a continuous experiment by decrementing the number of timing pulses logarithmically until the upper limit of frequency set by the operator is reached. During the variation of driving frequency, either the shear-strain amplitude θ_o or the shear-rate amplitude $\dot{\gamma}_o$ will be kept constant, as the operator has chosen. In the constant θ_o mode, the peak displacement of the outer cylinder is constant; since $\theta_o = A/(2\pi f K_V)$, the peak drive voltage A is increased linearly with increasing frequency. In the constant $\dot{\gamma}_o$ mode, since the amplitude

$$\dot{\gamma}_o = 509\theta_o f = \frac{509}{2\pi K_V} A \tag{3}$$

the peak drive voltage A is held constant throughout the whole range of frequencies.

c. Superposition of Steady Rotation. The dynamic properties of a nonlinear fluid under pure oscillatory shear may be influenced by a superimposed steady-shear flow; and, in the study of time dependency, the buildup and breakdown times may depend on shear stress or shear rate. It is thus desirable to be able to superpose steady rotation on oscillatory shear. This can be easily accomplished by adding a selected constant to every point of the cycle of the sine function stored in memory, thus displacing the sinusoidal control signal by a fixed voltage, so that a combined steady and oscillatory shear of the servomotor is produced.

2. Triangle Wave

The hysteresis loop generation program provides means for increasing and decreasing the applied shear rates as a linear function of time. These controls are important where flow properties depend on time as well as shear rate. In this program, one cycle of a triangle wave control signal with desired period and amplitude is readily approximated by adding constant increments or decrements to the control signal at regular intervals. The increment or decrement is controlled by a two-digit octal number, 0 to 77. Since in the 12-bit DAC, a digital value 1 is converted to 2.44mV and 1 mV of control voltage in turn causes a rotational speed of 0.45 rpm of the motor, the increment or decrement of speed can be chosen in the range from 0.45 rpm to 28.2 rpm. By selecting the number of output pulses counted by the crystal clock, each rotational speed can be applied for a period ranging from 600 μsec to 157 sec. The maximum range of rotational speed this program can provide is up to 918 rpm, which corresponds to a shear rate of 7835 sec^{-1}.

D. The Response Signal--Data Acquisition and Analysis

1. Sinusoidally Oscillatory Shear

    a. Constant Frequency. The response signal from the torque transducer is periodic, either sinusoidal or nonsinusoidal, with the same frequency as the driving signal indicated by the tachometer. Both signals are sampled 32 times per cycle, i.e., with N output steps per cycle of the control signal, samples are taken every N/32 output steps through ADC. Sampling begins after a few cycles of oscillation, so as not to include starting transients, and continues for 8 successive cycles, after which the computer computes an average cycle for each signal by summing the corresponding digital data from each of the 8 cycles and then dividing them by 8. The averaged wave forms with 32 points each are then

displayed on the oscilloscope. An example of this is shown in Fig.
3. The wave forms can be displayed separately on command through a
switch on the computer's console. The digitized amplitudes of tachometer and transducer signal are further converted to the angular
displacement (in radians) of outer cylinder and inner cylinder respectively. The angular displacement data of inner cylinder are
then divided by the peak displacement value of the outer cylinder,
thus they are expressed in the form of "relative" angular displace-
ments.

To evaluate the amplitude ratio and phase angles of the response torque signal, both sets of data are Fourier-analyzed for the
first five harmonics as a cosine series:

$$f(t) = A_o + \sum_{n=1}^{5} A_n \cos \left(\frac{2n\pi t_f}{P} - \phi_n \right) \tag{4}$$

For a continuous time series, P is the period of oscillation and t_f
the time; for the discrete series as in this case, t_f is considered
the index of the point ($0 \leq t_f \leq 32$) and P = 32. A_n is the amplitude
and ϕ_n is the lag angle of the nth harmonic with reference to a pure
cosine wave of the same frequency $2n\pi/P$.

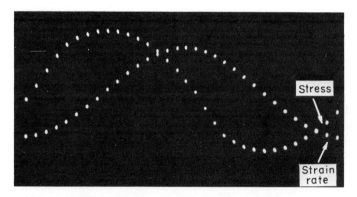

FIG. 3. Average cycle of stress signal and strain rate signal
as presented on oscilloscope.

Execution of the Fourier analysis requires only 4 sec after
which the results are displayed on the oscilloscope. Figure 4 is
a photograph of the alphanumeric display, the first column lists
the amplitudes and the second column the phases. Phase angles for
the response torque (channel 1) signal are referred to the phase of
the driving signal (channel 0). The bar-graph spectrum shown in
Fig. 5 is a comparison of amplitudes of different harmonics with
the largest amplitude taken as unity (a sinusoidal signal has only
the fundamental frequency). Options are provided to call forth
these different types of display of either channel. The alphanumeric results can be printed out through the teletype.

The last part of data analysis is to calculate the real, G',
and the imaginary, G''. components of the complex dynamic shear
modulus, G*, by using the amplitude and phase angle of the fundamental
frequency of the response stress data assuming the fluid is linear.
Therefore the G' and G'' values thus calculated are useful for
linear analysis only when the response stress signal shows negligible higher harmonic contents.

The oscillatory motion of a linear viscoelastic fluid in an
annular space between walls of two coaxial cylinders has been
studied in detail by Markovitz [8,9]. Giving the amplitude and

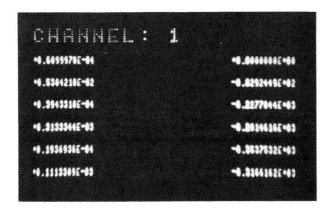

FIG. 4. Display of amplitudes and phase angles in Fourier
series for a stress signal.

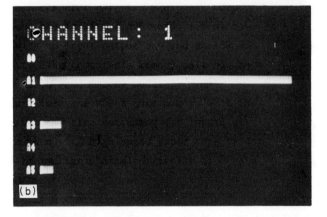

FIG. 5. Bar graph of spectrum of (a) a strain rate signal and (b) a nonlinear stress signal. Absence of harmonics in (a) indicates a pure sine wave.

frequency of the outer cylinder, he has shown that

$$1 - \frac{1}{m}\cos\phi - \frac{i}{m}\sin\phi + \frac{i}{\eta^*}[('A_1 + B_1\rho)\omega - \frac{C_1}{\omega}]$$

$$- \frac{1}{\eta^{*2}}[(A_2' + B_2\rho)\rho\omega^2 - C_2\rho]$$

$$- \frac{i}{\eta^{*3}}[(A_3' + B_3\rho)\omega^3\rho^2 - C_3\rho^2\omega]$$

$$+ \cdots = 0 \tag{5}$$

$$\eta^* = \frac{G^*}{i\omega} = \frac{1}{\omega}(G'' - iG') \tag{6}$$

where m is the amplitude ratio of the displacement of the inner cylinder to that of the outer cylinder; ϕ is the phase angle by which the inner cylinder lags the outer cylinder; ρ is the density of the test fluid; ω is $2\pi f$; f the frequency; η^* is the complex dynamic viscosity. A_n', B_n', and C_n (n = 1, 2, 3, ...) are geometric constants of the apparatus. Their first two terms are

$$A_1' = \frac{I_o(R_o^2 - R_i^2)}{4\pi L\, R_o^2 R_i^2}$$

$$B_1 = \frac{(R_o^2 - R_1^2)^2}{8\, R_o^2}$$

$$C_1 = \frac{K_\theta A_1'}{I_o}$$

$$A_2' = \frac{I_o}{32\pi L}\left(4\ln\frac{R_i}{R_o} + \frac{R_o^2}{R_i^2} - \frac{R_i^2}{R_o^2}\right) \tag{7}$$

$$B_2 = [(R_o^2 - R_i^2)(R_o^4 - 5R_o^2 R_i^2 - 2R_i^4)$$

$$+ 12R_o^2 R_i^4 \ln\frac{R_o}{R_i}]/192\, R_o^2$$

$$C_2 = \frac{K_\theta A_2'}{I_o}$$

where R_o and R_i are the radii of outer cylinder and inner cylinder, respectively, and L is the immersed length of inner cylinder. In our case we found

$$A_1' = 6.653 \times 10^{-2} \qquad A_2' = 3.040 \times 10^{-5}$$

$$B_1 = 7.779 \times 10^{-4} \qquad B_2 = 4.032 \times 10^{-6}$$

The terms with indices higher than $n = 3$ were all negligible. Therefore Eq. (5) can be solved for $1/\eta^*$ and thence η^* by use of the usual quadratic formula. The results are shown as follows: Let

$$(A_1' + B_1\rho) - \frac{C_1}{\omega} = D_1$$

$$(A_2' + B_2\rho)\rho\omega^2 - C_2\rho = D_2$$

$$1 - \frac{\cos\phi}{m} = D_4$$

$$\frac{\sin\phi}{m} = D_5$$

1. Assume $(D_1^2 - 4D_2D_4)$ is positive, then

$$G' = \frac{2D_2\omega(D_1 + r_A^{1/2}\cos\frac{\theta_1}{2})}{2D_1 r_A^{1/2}\cos\frac{\theta_1}{2} + D_1^2 + r_A} \tag{8}$$

$$G'' = \frac{-2D_2^\omega r_A^{1/2}\sin\frac{\theta_1}{2}}{2D_1 r_A^{1/2}\cos\frac{\theta_1}{2} + D_1^2 + r_A} \tag{9}$$

where

$$r_A = \left[(D_1^2 - 4D_2D_4)^2 + (4D_2D_5)^2\right]^{1/2} \tag{10}$$

and

$$\theta_1 = \tan^{-1}\left[\frac{4D_2D_5}{D_1^2 - 4D_2D_4}\right] \tag{11}$$

If G' and/or G'' are negative, θ_1 is replaced by $\theta_1 + 2\pi$ in Eqs. (8) and (9) for G' and G''.

2. If $D_1^2 - 4D_2D_4 < 0$, θ_1 is replaced by $\theta_1 + \pi$ in Eqs. (8) and (9) for G' and G''/ The computer completes the experiment by printing out the G' and G'' values through the teletype.

b. *Frequency Scanning*. When the system is carrying out a frequency-scanning experiment, at each driving frequency the same operations of data acquisition and analysis as described in Sec. II.D.1.a are performed, except for the oscilloscope displays. As the experiment continues, the computer stores the G' and G'' data as a function of frequency; after it has scanned through the whole range of frequencies, $\log_{10}G'$ and $\log_{10}G''$ are plotted as functions of $\log_{10}\omega$ by the X-Y reorder. An example of this is shown in Fig. 6. A similar plot with labeled scales is shown in Fig. 9.

2. Hysteresis Experiment

Multiple samplings of the tachometer signal and the torque transducer signal are taken and averaged at each speed level. Since the phase angle between these two signals is not important in this kind of experiment, the unattenuated tachometer signal is plugged directly into the ADC instead of passing through the twin filter unit. Conversion factors between the tachometer output voltage and the rotational speed, previously obtained by calibration, were entered in the program for computation.

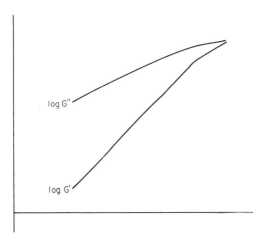

FIG. 6. Original drawing by the X-Y recorder in an oscillatory shear experiment. The abscissa is log ω.

After the speed has returned to zero, the data are converted
to shear stress (τ) and apparent shear rate ($\dot{\gamma}_a$) as follows:

$$\tau = \frac{d K_T}{2 R_i^2 L} \tag{12}$$

$$\dot{\gamma}_a = \frac{\pi U}{15 \left(1 - \dfrac{R_i^2}{R_o^2}\right)} \tag{13}$$

where d is the deflection of inner cylinder in mil; U is the rotational speed in rpm. The dimensions of two cylinders R_i, R_o, and L
are in cm. The data can be printed out on the teletype together
with the apparent viscosities ($\eta_a = \tau/\dot{\gamma}_a$). They can be plotted on
the X-Y recorder with shear stress as a function of apparent shear
rate; a hysteresis loop is formed if the test fluid has time-dependent behavior. Some examples are shown in Figs. 10, 11, and 12.

To characterize steady-state flow behavior, the desired shear-
rate range is scanned in discrete steps using this program, at a
rate sufficiently slow that the upcurve and downcurve coincide.

III. DISCUSSIONS OF PRELIMINARY RESULTS

To examine the ability of this instrument, extensive sets of
data were collected for different types of fluid. The results are
compared with theoretical predictions or results obtained by other
research groups.

A. Oscillatory Shear Experiments

1 Newtonian Fluid

If a Newtonian fluid is put into the rheometer, η^* will be a pure
real number. In this case, Eq. (6) can be separated into real and
imaginary parts. The resulting equations, considering terms up to
those of order $(1/\eta^{*2})$, are then

$$\left[(A_1' + B_1\rho)\omega - \frac{C_1}{\omega} \right]\frac{1}{\eta} - \frac{\sin\phi}{m} = 0 \tag{14}$$

$$- \left[(A_2^{'} + B_2\rho)\rho\omega^2 - C_2\rho \right] \frac{1}{\eta^2} + 1 - \frac{\cos\phi}{m} = 0 \qquad (15)$$

As a result, the plots of (f sin ϕ/m) and 1 - (cos ϕ/m) as a function of f^2 should give straight-lines whose slopes and intercepts are determined by Eqs. (14) and (15).

The Newtonian fluid used was Amoco Chemical Co. "Indopol" Polybutene L-100, of viscosity 3.183 poise as determined previously using an Ostwald viscometer at 30.0°C. A series of tests was performed for this fluid at 30.0°C using different torsion bars. A typical result is given in Fig. 7. In this particular example, the 1/16 in. torsion bar was employed. The open circles represent the experimental points obtained for pure oscillation. The solid lines are calculated from the instrumental dimensions and constants and the known viscosity of the oil. The agreement between the line calculated from the theory and the experimental points is considered to be satisfactory. In the plot of 1 - (cos ϕ/m) as a function of f^2, some scattering occurs in the lower frequency region. In general, a series of torsion bars with widely differing torsion constants was needed to cover a wide range of frequencies; this result indicates that a thinner torsion bar is needed for lower frequencies.

The G' and G'' values typed out by the computer are also examined and plotted as a function of frequency. As shown in Fig. 8, G' increases linearly with the frequency with a slope of 1.0 on the logarithmic plot. The proportionality constant between G'' and the angular frequency ω is 3.400, which is close to the viscosity of this oil. The G' values are negligibly small and not in the range covered by this graph. In theory for the case of a purely viscous fluid with a constant viscosity η, the shear storage modulus G' should be identically zero irrespective of the frequency, and the shear loss modulus G'' should be given by $\eta\omega$. Our experimental observations are in good agreement with the theory.

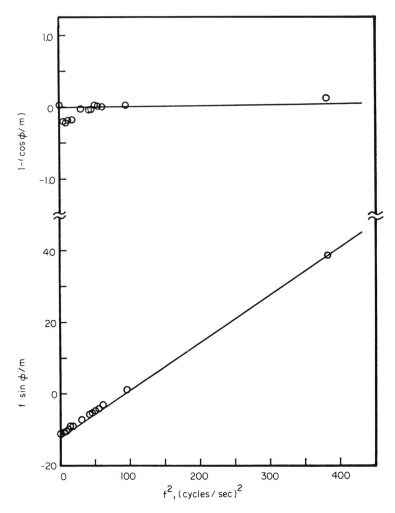

FIG. 7. f sin ϕ/m and 1 - (cos ϕ/m) plotted as functions of
f² for L-100 oil at 30°C. The solid lines are calculated by Eqs.
(14) and (15). The circles are experimental points.

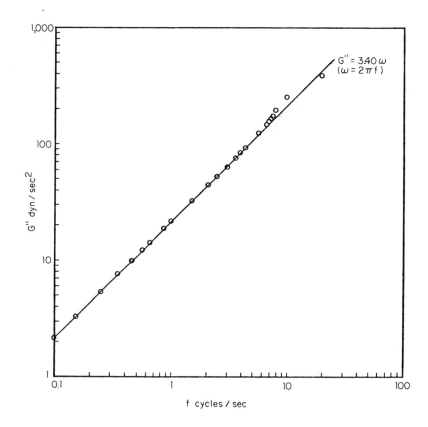

FIG. 8.   Frequency dependence of shear loss modulus of L-100
oil at 30°C.

2. Viscoelastic Fluid

A viscoelastic fluid was also tested to determine its moduli, G'
and G''. as a function of driving frequency in the oscillatory experiment. The sample labeled NBS Nonlinear Test Fluid No. 1 was
obtained through the kindness of Dr. Elliot A. Kearsley of the
National Bureau of Standards. It was a 6% solution of polystyrene.
Along with the sample, dynamic shear data collected by several
other research groups at 25°C were also supplied; they are represented by the two curves in Fig. 9.

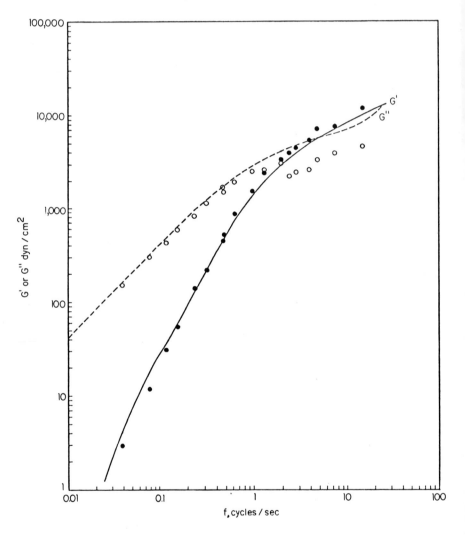

FIG. 9. Frequency dependence of dynamic shear moduli of NBS
Nonlinear Test Fluid No. 1. The lines are NBS experimental data.
The circles are our experimental data: (●)G' and (o)G''.

Due to the lack of a cooling system, the water temperature in our constant temperature bath was maintained at 26.0°C, which is 1°C higher than the room temperature when the measurement was made. The 1/8 in. torsion bar was used in the experiment. The amplitude of oscillation of the outer cylinder was kept so small that the higher harmonic contents of oscillation of the inner cylinder were negligible throughout the experiment. Three different strain amplitudes (0.0015, 0.003, and 0.007 radians) were employed; the variations did not cause significant change in the resultant data of amplitude ratio and phase angle. It appears therefore that the strain amplitude was sufficiently low for linear viscoelasticity experiments.

In Fig. 9, the closed (G') and open (G'') circles represent our experimental data. All values given are averages of several measurements at the same frequency and strain amplitude. The agreement between the data obtained by our instrument and those furnished by NBS in the same temperature range is considered to be reasonably good.

B. Hysteresis Loop Experiments

1. Newtonian Fluid

The L-100 oil was tested with the hysteresis loop program to insure that the equipment or experimental technique did not produce hysteresis loops. A series of tests of various up-down cycle periods over the same shear-rate range at 30.0°C were performed. The flow curves drawn by the X-Y recorder are shown in Fig. 10. The up curve (shear rate increasing) and the down curve (shear rate decreasing) are indicated by arrows. For the purpose of comparison, the scales of abscissa or ordinate of different figures are consistent. Part (a) is the result of steady-state shear; both the up and the down curves of this longest cycle period are essentially coincident straight lines. The slope of straight line gives the viscosity as $\eta = 3.400$. With the other tests of long cycle period, (b) and (c), the sample still showed Newtonian behavior. However, when the

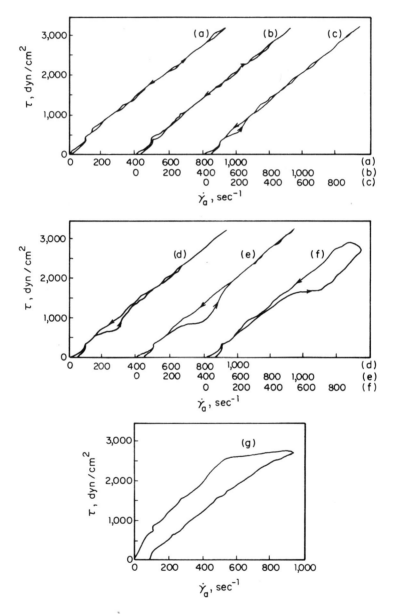

FIG. 10.   Flow curves of seven different up-down cycle periods for L-100 oil at 30°C.   (a) 963.4 sec, (b) 51.6 sec, (c) 25.8 sec, (d) 12.9 sec, (e) 6.5 sec, (f) 1.6 sec, (g) 0.5 sec.

period was reduced to 12.9 sec, a loop started to form; the size of the loop became more significant with faster cycling speed.

The hysteresis loops generated in the short period experiments indicate that the viscosity values of the down curve are higher than those of the up curve. This is the evidence that the loops were not caused by frictional heating in the fluid since temperature rise lowers viscosity. One fact which may explain the loop generating is that the transducer response signal has to go through a filter unit before it is measured by the ADC, and the filter unit has been observed to cause phase shift as a signal was fed in. This time lag between the input and output signals would be measured all along the loop in the hysteresis-loop experiment. It would especially result in significant loop size in tests of very short cycling periods for Newtonian fluid since the phase shift caused by the filter unit was found to increase with increasing frequency when a sinusoidal signal was fed into it.

The observations in this experiment suggest that to carry out a hysteresis-loop experiment with this program, the up-down cycle period should be at least 15 sec. Therefore this type of experiment is not suitable for fluids of very short characteristic time.

2. Thixotropic Fluid

The thixotropic fluid used was 9% bentonite aqueous suspension. The bentonite powder was purchased from the Fisher Scientific Company, mixed with water, and stirred by a blender for 1 hr. The tests were carried out at 30.0°C at three different cycling periods: 30.1, 60.0, and 120.2 sec; all are with the same peak shear rate of about 325 sec^{-1}. The sample was allowed to rest for 40 minutes before the test of another cycle period was started. The flow curves are shown in Fig. 11. Thixotropy is indicated by the fact that the shear stresses on the up curve are higher than the corresponding ones on the down curve, indicative of a gradual breakdown of thixotropic structure on the up curve and a gradual rebuilding on the down curve. The bulge on the up curve at low shear rates is another feature of thixotropy, which was also observed by other research

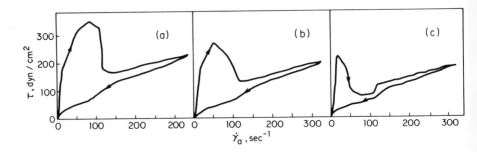

FIG. 11. Hysteresis loops of three different up-down cycle periods for 9% (by weight) bentonite aqueous suspension at 30°C. (a) 30.1 sec, (b) 60.1 sec, (c) 120.2 sec.

groups [10,11,12,13] with the same type of suspension. In this range the structural breakdown under shear has a greater influence on the shear stress than the increasing shear rate has. The size of the bulge is smaller for the test of longer cycle period; this is due to the fact that longer times were allowed at each shear rate, so that the response stress can reach closer to the equilibrium state. Also illustrated in this figure is another feature which is common to probably all thixotropic fluids; the response stress recorded at the peak shear rate increase with increasing cycle speed.

The extent of thixotropy of different substances can be measured by comparing the size of hysteresis loops when similar controlled shear-rate variations are applied to them, as the speed of restoration of structure is a major factor in determining the size of the area enclosed in the loop.

C. Measurement of Time Dependency by Oscillatory Shear

To measure the time dependency of the viscosity of a viscous fluid, data obtained from this instrument are analyzed using the technique developed by Dodge and Krieger [5,6]. It consists of several steps. First, the steady-state data of apparent shear rate as a function of shear stress are obtained, which corresponds to the equilibrium flow curve of the fluid. Their computer program, "VISFIT," is used for the UNIVAC 1108 computer to fit these data, determines the true viscosity as a function of true shear rate, and fits the following Ree-Eyring [14] equation to the resultant data:

$$\eta = a + b \, \frac{\sinh^{-1}(\beta_1 \dot{\gamma})}{\beta_1 \dot{\gamma}} + \frac{\sinh^{-1}(\beta_2 \dot{\gamma})}{\beta_2 \dot{\gamma}} \tag{16}$$

Another computer program, "NONCYL," written by the same authors is used to solve by finite difference technique based on the steady-state behavior and to predict the oscillatory shearing behavior of the fluid. Next, an oscillatory experiment scanning a wide range of frequencies with constant shear rate amplitude is carried out. At low frequencies, where the response time of the fluid is short compared to the period of oscillation, the observed and predicted data should be in agreement. At frequencies where the time dependency of the viscosity is important, there will be discrepancies between the experimental and the predicted results.

Steady-state and oscillatory shear measurements were made for a non-Newtonian fluid at 30.0°C. The sample, Latex MJ72B, was a uniform polystyrene latex with particles of diameter 9040 Å prepared in our laboratory by emulsion polymerization and concentrated by rotary evaporation under reduced pressure to a solid content of 46.4% by volume. It has sufficient electrolyte added to be at the minimum of viscosity as a function of electrolyte concentration.

Oscillatory data were taken for several different frequencies ranging from 0.0781 to 13.02 Hz using the 1/16 in. torsion bar. The shear-rate amplitude was held at 25.77 sec^{-1}, which is in the non-Newtonian region as observed from steady-state measurement. For the purpose of comparison, the two sets of Fourier coefficients for each frequency, i.e., rheometer-measured and computer-predicted, are used to resynthesize the waveforms using the UNIVAC computer. Some sample plotter outputs are shown in Figs. 12 to 14. For this latex sample it was found that below 1.302 Hz the agreement between observed and predicted data are very good, both in amplitude and

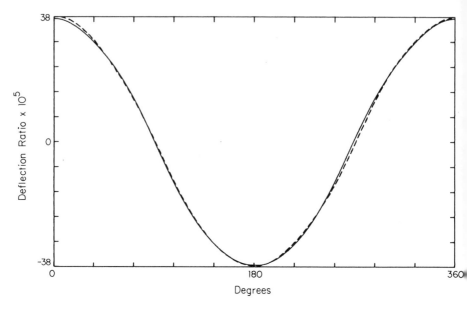

FIG. 12.   Comparison of predicted (solid) and experimental
(dashed) results for oscillatory shear at 0.156 cycles/sec, 0.463
rpm superposed rotation for latex MJ72B.

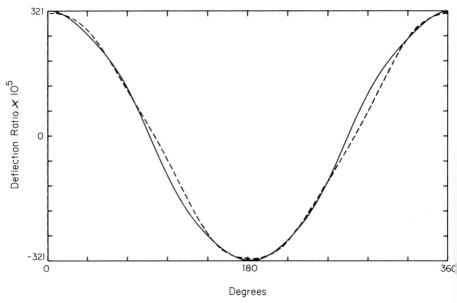

FIG. 13.   Comparison of predicted (solid) and experimental
(dashed) results for oscillatory shear at 1.302 cycles/sec, 0.340
rpm superposed rotation for latex MJ72B.

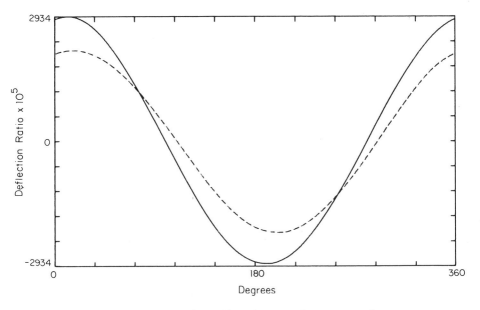

FIG. 14.   Comparison of predicted (solid) and experimental
(dashed) results for oscillatory shear at 6.51 cycles/sec, 0.317
rpm superposed rotation for latex MJ72B.

phase, whereas above this frequency, the lack of agreement is significant. This indicates the response time of the fluid is of the
order of one second.

IV.   SUMMARY AND CONCLUSIONS

A new rheometer was constructed and interfaced to a minicomputer system. This instrument has been demonstrated to have the
capability of performing steady-shear, hysteresis-loop generation,
and oscillatory shearing experiments, as demonstrated by experimental data obtained for (1) Newtonian, (2) thixotropic, (3) viscoelastic, and (4) nonlinear viscous fluids. It satisfies most of
the instrumental needs for studying nonlinear and time-dependent
fluids with an oscillatory shear program providing a wide frequency
range, large shear amplitudes at low frequencies, good signal-to-
noise ratio, automatic frequency scanning, and immediate data analysis. Using the computer to generate the control signal, the

frequency can be varied in small steps with an upper limit of about 156 Hz. Also the hysteresis-loop generation program provides a fast and convenient method to compare the degrees of thixotropy of different fluids. The prompt presentation of results from both programs allows the experimenter to decide whether the experiment need be repeated or to modify the experimental conditions such as amplitude and frequency in the oscillatory experiment or shear-rate range and experiment time in the hysteresis-loop experiment.

APPENDIX: SYMBOLS AND ABBREVIATIONS

A	Amplitude factor
A_n	Amplitude of the nth harmonic in the Fourier series
ADC	Analog-to-digital converter
BIN	Binary loader
d	Deflection of inner cylinder in mils
dc	Direct current
DAC	Digital-to-analog converter
f	Frequency in cycles/sec
f_L	Starting frequency in frequency scanning program
f_n	Highest frequency in frequency scanning program
G^*	Complex dynamic shear modulus
G'	Shear storage modulus, the real component of G^*
G''	Shear loss modulus, the imaginary component of G^*
I_o	Moment of inertia of the inner cylinder assembly $(g \cdot cm^2)$
K_θ	Torsion bar constant in dyne·cm/radian
K_T	Torsion bar constant in dyne·cm/mil
K_v	The ratio of motor speed to control voltage in V/rad/sec
L	Immersed length of inner cylinder
M	Number of clock pulses between output steps
m	Amplitude ratio of the displacement of the inner cylinder to that of the outer cylinder
N	Number of sine values stored in computer for one cycle of output driving signal
P	Period of oscillation in the Fourier series

R_o Radius of outer cylinder

R_i Radius of inner cylinder

rpm   Revolutions per minute

SR   Switch register

t   Time in seconds

t_f Index of point in the averaged waveform

U   Rotational speed of outer cylinder in rpm

V_θ Angular speed of motor in radians/sec

$\dot{\gamma}_o$ Shear-rate amplitude in sec^{-1}

$\dot{\gamma}_a$ Apparent shear rate

η Viscosity in poise

η^* Complex dynamic viscosity

η_a Apparent viscosity

ω Angular frequency in radians/sec

Ψ_i The angle 2 i/N, where i = 0, 1, 2, ..., N

ϕ_n Lag angle of the nth harmonic in the Fourier series

ϕ Phase angle by which the inner cylinder lags the outer cylinder

ρ Density

τ Shear stress

Θ The displacement of motor or outer cylinder in radians

Θ_o Strain amplitude of oscillation of outer cylinder in radians

REFERENCES

1. D. C. H. Cheng and F. Evans, *Brit. J. Appl. Phys.*, *16*, 1599 (1965).

2. J. Harris, *Nature*, (London), *207*, 744 (1965).

3. H. Green and R. N. Weltmann, *Ind. Eng. Chem.*, *Anal. Ed.*, *18*, 167 (1946).

4. J. Harris, *Nature*, (London), *214*, 796 (1967).

5. J. S. Dodge, Ph.D. Dissertation, Case Western Reserve University, Cleveland, Ohio, 1969.

6. J. S. Dodge and I. M. Krieger, *Trans. Soc. Rheol.*, *15:4*, 589 (1971).

7. *The Weissenberg Rheogoniometer Instruction Manual*, Farol Research Engineers, Ltd., Bognor Regis, Sussex, England.

8. H. Markovitz, *J. Appl. Phys.*, *23*, No. 10, 1070 (1952).

9. H. Markovitz, O. M. Yavorsky, R. C. Harper, Jr., L. J. Zapas, and T. W. DeWitt, *Rev. Sci. Instr.*, *23*, 430 (1952).

10. K. Norrish, *Discussions Faraday Soc.*, No. 18, 1954, pp. 120-134.

11. W. F. Gabrysh and H. Eyring, *J. Am. Ceram. Soc.*, *46*, No. 11, 523 (1963).

12. P. Rebinder, *Discussions Faraday Soc.*, No. 18, 1954, pp. 151-160.

13. F. Moore, *Trans. Brit. Ceram. Soc.*, *58*,(7-8), 470-494 (1959).

14. T. Ree and H. Eyring, *J. Appl. Phys.*, *26*, 793, 800 (1955).

15. T. F. Niu, Ph.D. Dissertation, Case Western Reserve University, Cleveland, Ohio, 1973.

16. G. I. Taylor, *Phil. Trans. Roy. Soc.*, (London), *A223*, 289 (1923).

Chapter 8

COMPUTER TECHNIQUES FOR KINETIC STUDIES IN
THERMAL ANALYSIS AND RADIATION CHEMISTRY OF HIGH POLYMERS

Walter Y. Wen

Plastics Laboratory
Honeywell Inc.
Hopkins, Minnesota

Malcolm Dole

Department of Chemistry
Baylor University
Waco, Texas

I. INTRODUCTION

During the past 10 years, computer technology has developed to
the point where virtually all areas of science, engineering, and
business have shared in its success. At first, the high-speed
electronic computer was used simply as a remote after-the-fact
"super-adding machine" for complicated computations; without this
super machine, solutions to these problems were practically inac-
cessible.

In 1963, a relatively new type of small digital computer was
introduced, a type that revolutionized the concept of laboratory
instrumentation. The art of science and technology has advanced to
such a sophisticated state that no human errors can be tolerated;
therefore, computer-controlled analytical devices have become a
matter of necessity. Low cost, compact computers, known as minicomputers have demonstrated their distinctive abilities in the area
of laboratory instrumentation when compared with skillful operators
running different types of experiments without computers. Recently,
the art of computerized instrumentation has advanced to where the
minicomputer can be used in production lines in various industries.
Polymer chemists have used electronic computers as super-desk cal-
culators for a long time, but attempts at employing minicomputers
in their experimental work did not start until recently.

Although applications of computer techniques are extremely
broad, all uses can be logically classified into two categories:
instrumentation and computation. These two types of computer applications are obviously different, but their techniques and approaches are compatible. This chapter is organized according to
these two categories. Because of the rapid growth and volume of
literature accumulated in this area, it would be very difficult to
study extensively all related topics. Attention is concentrated
only on the subjects directly concerned with the authors' areas of
interest and experience.

In this chapter, we review the computer techniques used in our
laboratories for data collection, and for the interpretation of

data obtained in thermal analysis studies and in experiments on the radiation chemistry of high polymers. The first two sections describe computerization of the instrument involved using two different types of small computers. At the Honeywell Plastics Laboratory, a minicomputer, Honeywell H112, has been interfaced with the Thermogravimetric Analysis (TGA) and a Differential Scanning Calorimeter (DSC). A software pack has been developed specially for this instrument. Under the computer control mode, data axquisition is totally automatic. A general discussion of the software techniques with this system is given. At the Baylor University laboratory, a different minicomputer, Varian 620/i, was coupled with an electron spin resonance (esr) spectrometer for data acquisition and reduction in the study of gamma radiation effects in high polymers. Software for operating this machine was provided by the vendor of the computer. A brief description of the features and options of the program is given.

In the area of data interpretation, mathematical expressions for TGA and DSC kinetics have been derived. These expressions are employed in a FORTRAN program that enables the calculation of the kinetic parameters of an anticipated reaction order from a single thermogram of a temperature-programmed experiment. The program was designed mainly for first-order degradation processes, but it can handle reactions of different reaction orders as well. A general differential technique, processed on a timesharing device of a Honeywell Computer Network (HCN), has been adopted for the program. Data from esr experiments are treated to test the theoretical equations describing the first-order alkyl readical decay or the diffusion-controlled second-order decay of allyl radicals in polyethylene (PE). FORTRAN programs, FORTRAN II and IV, were prepared for data processing on a Honeywell 1620 computer at Baylor University or on a CDC 6600 terminal from the University of Texas at Austin.

A section of the chapter is devoted to computer application in analysis of sol-gel data of irradiated PE according to the Wesslau molecular weight distribution function. The chapter ends with

plans for future development of minicomputers for laboratory automation at the Plastics Laboratory within Honeywell.

II. MINICOMPUTERS IN INSTRUMENTATION

Electronic computers usually fall into two categories: analog
and digital. In some instances, the execution speed of an analog
computer is faster than that of a digital computer of the same
class [1]. However, an analog computer is not suitable for use in
computerizing instrumentation because of its inflexibility. Usually an analog-computer-automated instrument is restricted to a particular purpose and a change of the monitoring program has to be
done through the computer hardware (electronic circuitry), which
makes program debugging processes extremely tedious and time-
consuming, if not impossible. Actually, the idea of computer-
controlled instrumentation was not substantiated until the introduction of the digital minicomputers. As the name implies, a minicomputer is a small computer that has limited capability even though
it has essentially the same features as any computer in its class.
Because of the compact size and low price, minicomputers are most
desirable for applications in laboratory instrumentation [2-8] and
field operations [9-11]. A great majority of the minicomputers
produced have been applied directly or indirectly to monitoring use.
Today, it is clear that a large-scale automatic-control system can
be established using minicomputers as the building blocks [12-14].

There are over 200 different models of minicomputers on the
market in the United States [15], and most of them can be applied
to instrumentation purposes. Here, attention is focused on two independent types of minicomputers, a Honeywell H112 and a Varian
620/i, and their uses in computer-controlled thermal analysis and
esr experiments on polymeric materials are demonstrated.

A. Programming Techniques for Minicomputers

Almost all electronic computers operate on two signals, i.e.,
switch on or off, or in machine language binary code 0 or 1 [8,16,
17]. In general, a minicomputer contains 12 to 16 of the on-off

switches or, in computer terminology, bits. A sequence of computer
actions is actually the combined result of a computer's hardware,
which has incorporated in it many of these carefully programmed on-
off signals or binary codes. Unfortunately, every computer manufacturer has his own preference of hardware methods and this directly affects the language coding; as a result, programming techniques
(software) for all levels of language can be different from one
computer to another. Today, scientists of various areas are familiar with one or several high-level computer languages because of
their needs in complex data processing. However, the techniques of
programming a minicomputer are significantly different from high-
level language methods. These techniques involve some knowledge of
computer hardware, which apparently is not attractive to most chemists.

A computer capable of handling a high-level language, e.g.,
BASIC, FORTRAN, ALGOL, COBOL, or PL/I, must be equipped with a software package known as an assembler or compiler of that language.
The assembler or compiler assembles or compiles the instructions of
a program and translates them into the binary code machine language
that the computer "understands." The assembler or compiler acts as
an interpreter between the program and the computer. In a minicomputer, there is not enough room in the computer memory to accom-
odate an assembler or a compiler.

Sometimes a program can be assembled in a large computer and
the assembled object program is then input into the minicomputer.
This programming technique can be used only if an assembler that is
compatible with the hardware of both the large and the small computers is available. Thus, it is obvious that the programming
technique depends very much on the availability of the large computer. Very often, the vendor of minicomputer supplies a highly
efficient utility program that can serve as an assembler for a low-
level assembly language such as a mnemonic binary code language.
In this case, the program assembly process usually is very slow and
proceeds line by line. On the other hand, work can be done directly

on the binary code machine language although it would be extremely
difficult and even impossible to do in some complex programs.

It has been noted recently that high-level languages such as
BASIC and FORTRAN can be employed on real-time minicomputers if
they are equipped with proper peripheral devices and library [4,5].
In general, a programmer's manual is still the best reference for
details of programming techniques even though some beginners might
find them difficult to understand. For fundamental knowledge in
binary code machine language, mnemonic and assembly language, the
reader is referred to the literature [16,17].

In the following sections the authors' efforts to computerize
thermal analysis and esr equipment in their laboratories are explained.

B. The Dedicated Computer in Thermal Analysis

The Honeywell H112 minicomputer was selected for the present
study because this type of computer was originally designed for applications in real-time control, on-line, data acquisition and data
reduction. Because of the small capacity of the H112 system, the
task of software development in our experiments requires the knowledge of a low-level assembly language. Accordingly, the services
of an experienced programmer in this area are needed. Before going
into details of the system programming techniques, it is necessary
to give in a logical order a brief but clear description of the
associated system hardware.

1. Hardware

The computer hardware used was completely contained within the H112
system, which is shown in block diagram format in Fig. 1. The hardware package, system interface, X-Y plotter, and oscilloscope were
mounted on a rack as shown in Fig. 2.

   a. Computer. The H112 is a 12-bit computer with a memory field
of 4K that is expandable to 8K. [1K equals 1024 decimal (1024_{10}) or
2000 octal (2000_8) number computer locations. The reader is referred to Refs. 16 and 17 for the octal number system and the relations

among the binary, octal, and decimal systems.] The system memory
cycle time is 1.69 μsec and the execution time for a typical instruction runs from 2.54 to 7.63 μsec. The computer operates on
two's complement number system.

   b. Interface. The H112 interface consists of a 12-bit analog-
to-digital converter (ADC) with 40 μsec conversion time, a two-
channel 12-bit digital-to-analog converter (DAC), a real time clock,
and 12 lines parallel output. A block diagram of this interface
system is shown in Fig. 3. The two-channel DAC is flexible in that
it can be expanded to a double-channel system by simply adding another ADC card. Both converters operate on two's complement notation. The clock is programable at a sampling rate of 1 msec to
4096 (2^{12}) sec per point.

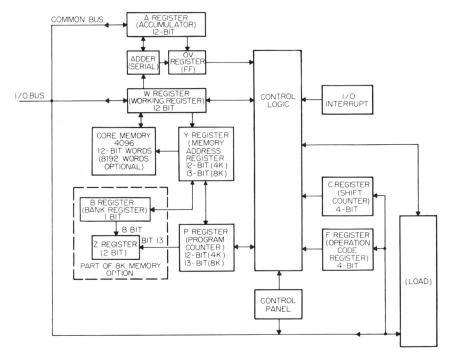

   FIG. 1. H112 controller-system block diagram. (Copyright
April, 1970, Honeywell Inc.; reprinted with permission of Honeywell
Inc.)

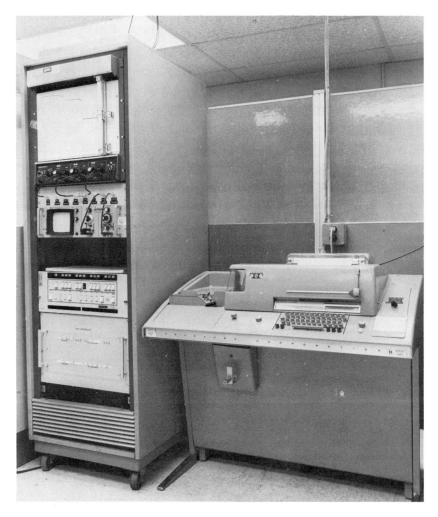

FIG. 2. H112 controller system.

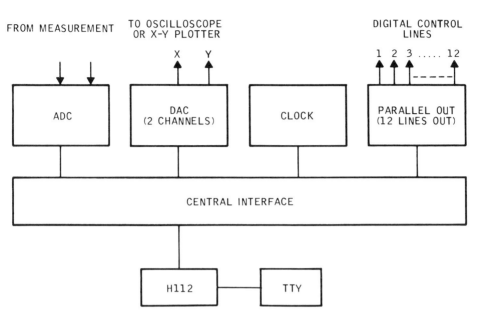

FIG. 3. H112 interface block diagram.

c. *Teletype*. An ASR-35 teletype was employed with the computer system. It was equipped with a 10-characters/sec paper tape
reader and punch. The teletype system is also shown in Fig. 2.

d. *Oscilloscope*. Data from the computer buffer can be di-
rected through the DAC and displayed rapidly on the oscilloscope
for visual convenience.

e. *X-Y Plotter*. The data also can be transmitted through the
DAC and traced on the plotter at a desired rate for later reference.

2. Software

Honeywell computer experts have developed a special assembly for the
H112 system. This language, known as System Assembly Programming 12
or SAP-12, consists of 37 standard instructions and 21 pseudo-operations. The pseudo-operations allow the execution of some special
functions that do not have counterparts in the machine language. A
SAP-12 source program has to be assembled in one of the Honeywell

Series 16 computers, i.e., H316, H516, and H716 using a compatible
SAP-12 assembler. The assembler assembles the program instructions
one line at a time according to the procedure given in Fig. 4. The
assembled object program is punched on an object tape that can be
read directly into the H112 for use.

Because a large computer of the Honeywell 16 family was not
immediately available at the time the software task was begun,

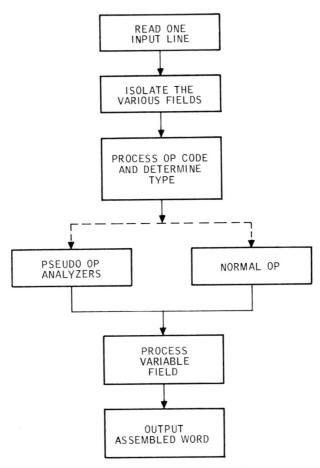

FIG. 4. Process of one line assembly. (Copyright April, 1970,
Honeywell Inc.; reprinted with permission of Honeywell Inc.)

however, we decided to proceed with our project using a low-level
mnemonic language. This language is just one level higher than the
machine coding. This language was used primarily because it could
be assembled by the Hll2 Utility Program known as Maxbug (MB).
Originally, the MB program was designed as a programing aid, which
makes possible a prompt communication with the computer central
processor and thus facilitates the program debugging operation.
Under the MB control mode, a source program can be entered through
the teletype one line at a time. MB assembles the instruction according to a process identical to that given in Fig. 4.

Our software package was developed step by step, usually following the order given in Fig. 5. The necessary auxiliary routines
general to all major subroutines were first developed independently.
Among these routines were the binary-to-decimal, octal-to-binary,
teletype message, and several other housekeeping programs. The
major subroutines were selfcontained and independent of one and another. As shown in the block diagram of Fig. 6, all major subroutines were linked only to the main program (Executive or EXEC).

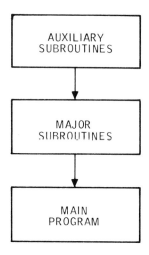

FIG. 5.   Procedure of software development for the Hll2.

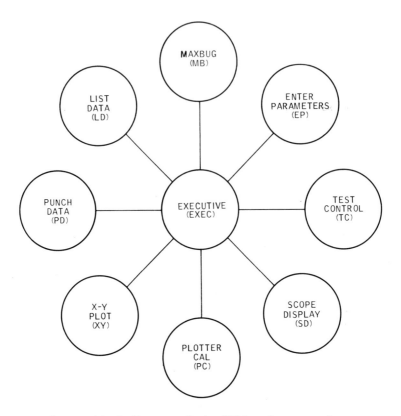

FIG. 6.   Block diagram of the H112 software package.

We believe that such an arrangement gives the software package maximum flexibility for future expansion of the system.   As indicated in Fig. 6, each major subroutine performs a particular task and the two-letter abbreviation of each subroutine is the notation used by the main program or EXEC to specify an operation.   The flow diagram of Fig. 7 shows that the EXEC is actually a programmed operation selector.   With this software package, a change of the experimental procedure can be readily established by modifying only the EXEC with all the subroutines unaltered.

The entire software program will not be explained, but an illustration of the programming techniques employed in the present study will be given.   Figure 8 is a block diagram and program list

of the scope display subroutines in our software package. The
first three columns of the list are memory location, machine coding,
and assembly language. The last column is the comments given by
our programmer for ease of inspection.

3. Operation

Because the H112 controller does not have an automatic start device,
the software programs have to be initiated from the front panel;
then an operation can be selected using the MB utility program which
starts at memory location 6000*.

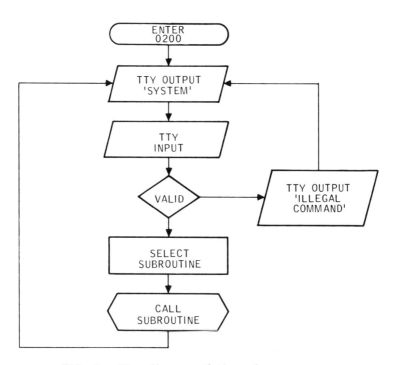

FIG. 7. Flow diagram of the main program.

---

*All computer memory locations are expressed in octal number unless
otherwise specified.

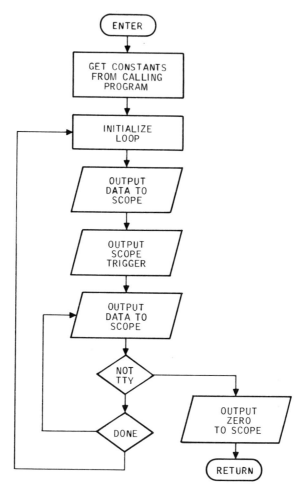

FIG. 8. Flow diagram and program list of the scope display
subroutine.

(continued)

```
**M2000,2077                           DSCOPE
02000  0200  NOP                       ENTER ON JST
02001  4600  LDA*                      GET DATA BUFFER START ADDRESS
02002  1270  STA                       TEMP STORE
02003  3200  IRS                       SET FOR BUFFER INDEX
02004  4600  LDA*                      GET BUFFER INDEX
02005  1271  STA                       TEMP STORE
02006  3200  IRS                       SET FOR RETURN
02007  0672  LDA                       GET BANK BIT
02010  0041  TAB                       SET BANK
02011  0670  LDA                       GET DATA BUFFER START ADDRESS
02012  1273  STA                       TEMP STORE FOR OUTPUT LOOP
02013  0671  LDA                       GET BUFFER INDEX
02014  1274  STA                       TEMP STORE FOR OUTPUT LOOP
02015  4673  LDA*                      GET FIRST DATA WORD
02016  4277  OTA                       OUTPUT TO SCOPE (PRESET)
02017  1616  JMP                       MAKE SURE
02020  0675  LDA                       GET TRIGGER BIT
02021  4236  OTA                       OUTPUT TO SCOPE EXT. TRIGGER
02022  1621  JMP                       MAKE SURE
02023  0060  CRA                       CLEAR A REG
02024  4236  OTA                       OUTPUT TO SCOPE EXT. TRIGGER
02025  1624  JMP                       MAKE SURE
02026  4673  LDA*                      GET FIRST DATA WORD
02027  4277  OTA                       OUTPUT TO SCOPE
02030  1627  JMP                       MAKE SURE
02031  4101  SKS                       CK FOR TTY INTERRUPTING
02032  1634  JMP                       SKIP TTY GET OUT
02033  1655  JMP                       GO RETURN TO CALLING PROGRAM
02034  0200  NOP
02035  3273  IRS                       INCREMENT DATA BUFFER ADDRESS
02037  1626  JMP                       GO OUTPUT NEXT DATA WORD
02040  4673  LDA*                      GET LAST DATA WORD
02041  4277  OTA                       OUTPUT TO SCOPE
02042  1641  JMP                       MAKE SURE
02043  0200  NOP
02044  0200  NOP
02045  1611  JMP                       GO REINITIALIZE AND START OVER
02046  0000  HLT
02047  0000  HLT
02050  0000  HLT
02051  0000  HLT
02052  0000  HLT
02053  0000  HLT
02054  0000  HLT
02055  0200  NOP                       ENTER ON TTY INTERRUPT
02056  0060  CRA                       CLEAR A REG
02057  0041  TAB                       RESET BANK BIT
```

(continued)

```
02060  4277  OTA                         OUTPUT ZERO TO SCOPE
02061  1660  JMP                         MAKE SURE
02062  5600  JMP*                        RETURN TO CALLING PROGRAM
02063  0000  HLT
02064  0000  HLT
02065  0000  HLT
02066  0000  HLT
02067  0000  HLT        CONSTANTS
02070  4000                              DATA BUFFER START ADDRESS
02071  6030                              BUFFER INDEX
02072  0001                              BANK BIT
02073  4000                              TEMP DATA BUFFER START ADDRESS
02074  6030                              TEMP BUFFER INDEX
02075  0001                              SCOPE TRIGGER
02076  0000
```

The description below follows the logical order of running an experiment, as shown in Fig. 6, going clockwise starting from the MB subroutine.

Maxbug (MB): The MB program is used to "turn on" the monitoring program and to change the content of any memory location. After the computer power is on and the teletype placed on-line, the operator clears all registers* and sets the P register to 6000, then depresses the "START" button. The computer responds with a carriage return (CR) and line feed (LF) then two asterisks (**) indicating that the computer is now under the MB control mode. This is the only occasion that the operator has to work from the front panel.

Sample Rate: Because the present system has a single channel ADC, it is applicable only to analyses involving measurement of a particular parameter as a function of time. The time variable is recorded by the clock shown in the block diagram of Fig. 3. The sample rate is determined by two numbers stored in address locations 1176 and 1177, respectively. The first number indicates the clock rate (RC) and the second number the sample interval (SI) based on the RC. Because of the limited memory locations in the H112 controller, the present software has not been programmed to enter a sample rate, both RC and SI, directly

---

* The definitions of memory registers are common to all computers. For details of H112 Controller, the reader is referred to the H112 Programmers Reference Manual, DOC. No. 701300722428, Order No. M-1164, April 1970, Honeywell Inc.

through the teletype. Instead, RC and SI are programmed to a measurement using the MB program. Only the 7 least significant bits (the right-most 7 bits) are used in the RC and only 1 of the 7 bits can be used at a time. The time intervals of the bits range from 1 μsec for the first bit to 1 sec for the 7th bit, with an increment of tenfold of a previous bit. The SI address is a full 12-bit word which must be entered in octal number. All inputs through the teletype are terminated by a CR. When SI contains 74_8 (60_{10}) and RC 100 (octal machine code*), the actual sample rate is one point per min.

To program the RC, the operator uses the input routine of the MB by typing "A1176" and a CR on the teletype. The computer responds by a LF and then types out the content of the memory location and halts. A machine code of four digits representing the desired RC can be entered. The program automatically steps to the next location, that is, 1177. The operator can repeat the input process if he wants to change the content in SI, or he can enter a CR to keep whatever is in the memory. A slash "/" will transfer the control back to the normal MB mode.

Obviously some, although not extensive, knowledge of octal number and machine code is needed to program the sample rate. However, with an understanding of the definition of machine code and with assistance of an octal-to-decimal table, operation ought to be straightforward. Nevertheless, we are modifying our software so that the sample rate can be input in normal decimal form through the teletype.†

Executive (EXEC): The main execution program is stored in the memory locations beginning at 200. To "turn on" the EXEC, the operator makes use of the MB subroutine to "jump" to location 200 by inputting "J200" and a CR on the teletype. The computer responds with a LF, prints out "SYSTEM?", a CR and LF, and comes to a halt. A subroutine can be selected by entering the associated two-letter notation. Commands other than those given in Fig. 6 will receive a message of "ILLEGAL CØMMAND!" as shown in the example in Fig. 9. When a subroutine execution is completed, the computer returns the control back to EXEC; a new operation can then be assigned.

Enter Parameter (EP): The number of data points intended to be collected in the measurement is entered under the EP subroutine. The computer types the message "INPUT NØ ØF SAMPLES!" and halts. As many as 1000 points can be stored in the computer memory during the measurement.

---

*Octal code is the grouping of 3 binary bits.

†The modification was completed after this manuscript was prepared.

```
**A1176
01176  0221        0100          ENTER RC
01177  0625        0074          ENTER SI
01200  1661        /             RETURN TØ MB
**J200                           GØ TØ EXEC

SYSTEM?
EP

INPUT NØ. ØF SAMPLES !1000

SYSTEM?
TC

TYPE S TØ START !S

SYSTEM?
SD

SYSTEM?
PC

SYSTEM?
XT
ILLEGAL CØMMAND!

SYSTEM?
PD

SYSTEM?
LD

SAMPLE NO.                     DATA VALUE

        0    2045    2047    2047    2047    2045    2047    2047
        7    2036    2041    2047    2047    2047    2045    2041
       14    2013    2029    2025    1984    2016    1967    1969

SYSTEM?
MB
**
```

FIG. 9. Examples of operating the H112 software programs.

Test Control (TC): This subroutine is nothing more than an
   immediate halt allowing the operator to adjust his instrument previously to the measurement. As shown in Fig. 9,
   replying to the statement "TYPE S TØ START!" with a "S"
   starts the data acquisition, which can be interrupted and
   terminated any moment by depressing the "/" key. Actually,
   all major subroutine operations can be terminated by
   entering the "/" on the keyboard. After the termination,
   the computer turns back to EXEC mode.
Scope Display (SD): At completion of the measurement, the
   data can be displayed on the oscilloscope. The trace can
   be erased from the scope by a "/".
Plotter Cal (PC): In case a hard copy of the trace is re-
   required for the record, the PC is selected for the operator to adjust the X-Y plotter scale so that the entire
   plot falls on the graph paper.
X-Y Plot (XY): This subroutine uses an algorithm of slew com-
   pensation, which puts a constant time delay on transfer-
   ring the variables to the computer buffer so that the pen
   speed of the plotter can catch up with the data conversion
   rate.
Punch Data (PD): The data can be punched on paper tape in
   decimal format for convenience of data processing.
List Data (LD): As shown in the example of Fig. 9, the data
   also can be printed on the teletype in decimal digits.
   Not more than eight numbers can be printed on a line.

4. Application

This computer system has been interfaced with a Cahn RG electrobal-
ance TGA or a Perkin-Elmer model 1B DSC to automate the associated
thermal analysis of constant heating rate experiments. Similar
computer-controlled systems for the TGA [18] and DSC [19] have been
reported recently.

Figure 10 shows the results of a TGA measurement on thermal
decomposition of polytetrafluoroethylene (DuPont TEFLON TFE) obtained by our computer-controlled system. The computer data have
been in good agreement with the corresponding trace obtained during
the experiment. The slight differences of the two curves can be
attributed to the difficulty of calibrating the X-Y plotter against
the graph paper and also to the slightly off-from-linear heating
rate due to rapid exothermicity at the peak of the decomposition.
However, in some cases, such as polyurethane foams, deviations of
the computer data from the experimental curve can be significant.

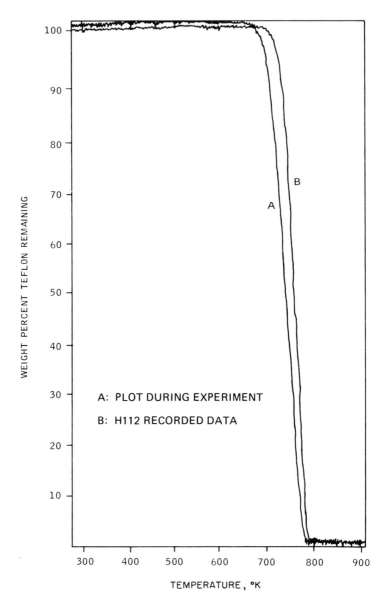

FIG. 10.   Thermal decomposition of TEFLON TFE as measured by
the H112 computer-controlled TGA.

Such deviations can be attributed to the nonlinearity of the heating rate due to the poor heat conductivities of the polyurethane
foams. A measurement on the curing of a styrene resin composite
using the DSC technique is given in Fig. 11.

There is no doubt that the present computer system can be
applied to experiments other than the temperature-programmed TGA
and DSC measurements. In fact, it can be coupled with any system
that measures a reaction variable as a function of time. Furthermore, the restriction of a "single variable as a function of time"
can be removed by adding another ADC card to the system interface.
Obviously, some minor changes in the software are necessary to
accommodate the dual channel input.

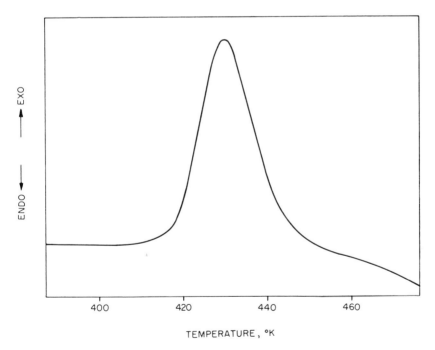

TEMPERATURE, °K

FIG. 11. Thermal decomposition of a styrene composite as
measured by the H112 computer-controlled DSC.

C. The Dedicated Computer in Electron Spin Resonance Experiments

The hardware and software of the esr system were purchased from Varian Associates. Although the Varian minicomputer requires a programing language completely different from SAP-12 and the low level mnemonic language used in the H112, the fundamental basis of the programing philosophy is essentially identical in both models of minicomputers. Because we do not have details of the program flow diagrams of the software package developed by Varian Associates, we cannot give a full discussion of the programing techniques. Instead, attention is concentrated on the operations of various options in the software package. We follow the same procedure that we have applied in describing the H112 system by starting our discussion with a very brief outline of the Varian system hardware.

1. Hardware

Computer hardware is the standard Varian product. The computer and system interface are enclosed in a cabinet shown in Fig. 12.

> Computer: The computer is a model 620/i with an 8K memory field expandable to 16K. It has a word length of 16 bits and a cycle time of 1.8 μsec.
> Teletype: An ASR-33 teletype equipped with a 10-characters/sec paper tape reader and punch is employed to communicate with the computer.
> Console: The SpectroSystem 100 (SS-100) is a keyboard controller. It has a programable digital display for a four-digit decimal number. The SS-100 console is also shown in Fig. 12. The console consists of a major keyboard, an operation and parameter board, and a message board as shown in Figs. 13, 14, and 15.
> Spectrometer: The spectrometer is an E-4 equipped with an oscilloscope and a X-Y plotter.

2. Software

The computer is programmed by an assembly language called DATA 620/i Assembly System (DAS). Details of programing techniques for DAS are given in the Computer and Programmers Reference Manuals[*].

---

[*]"Varian Data 620/i Computer Manual," Bulletin No. 605-A, Varian Data Machines; Programming Reference Manual, 1968.

FIG. 12.   The Varian 620/i computer system, the teletype, and
the SpectroSystem 100 console.

It has been demonstrated that an interactive language known as Conversational Language for Spectroscopic System (CLASS) can be effectively applied to the small core minicomputers [20].   CLASS is a
language that is easy to learn and understand.

Similar to the previous computer system, the Varian 620/i is
provided with an efficient utility program called AID, which performs

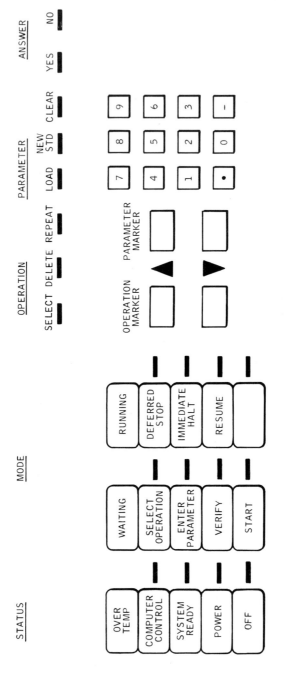

FIG. 13.   SpectroSystem 100 console keyboard.

MNEMONICS	OPERATIONS	MARK	PARAMETERS	MNEMONICS
SSCN	SINGLE SCAN		STANDARD	
MSAV	MULTI-SCAN AVERAGE		SCAN TIME (SEC)	TIME
PLDS	PLOT DURING SCAN		NUMBER OF SCANS	#SCN
DISP	SCOPE DISPLAY			
SMTH	SMOOTH		NUMBER OF POINTS	#PTS
SPNC	SPIN CONCEN-TRATION			
PLSS	PLOT STORED SPECTRUM		PLOT AMPLITUDE (% FULL SCALE)	PLAM
PLPS	PLOT PROCESSED SPECTRUM		SPECTRUM NUMBER	SPC#

FIG. 14. Operation and parameter board of the SpectroSystem 100 console for E-4.

NEED PARAMETERS	NEED COMPUTER CONTROL	CLEAR BEFORE MSA
SWEEP TOO SLOW	SWEEP TOO FAST	
	ADC OVER RANGE	SCAN TIME TOO SHORT
SCOPE FIELD SW OFF?	PEN OK?	DISPLAY NO. OF SCANS COMPLETED?
SUBTRACT?	PRINT VERIFIED PARAMETERS?	CLEAR SPECTRUM BEFORE MSA?
	SIGNAL AMPLITUDE TOO LARGE	
PARAMETER DISPLAY	SIGNAL AMPLITUDE DISPLAY	NUMBER OF SCANS COMPLETED

FIG. 15. Message board of the SpectroSystem 100 console for E-4.

the same tasks as MB does for H112. AID can be used to input experimental parameters and output experimental data on the teletype
or paper tapes. Because the 620/i memory field is twice as large
as that in the H112, it obviously can accommodate a much more sophisticated software package than that of the H112.

In addition to all the major subroutines similar to those in
the H112, the Varian software has eight independent programs that
perform different types of experiment or data reduction. The programs are Single Scan (SSCN), Multi-Scan Average (MSAV), Plot During
Scan (PLDS), Scope Display (DISP), Smooth Display (SMTH), Spin Concentration (SPNC), Plotted Stored Spectrum (PLSS), and Plot Processed Sprectrum (PLPS). The purpose of each program is clearly defined
in the names and the abbreviations of programs employed in the computer software. The SMTH and SPNC are data reduction routines; the
others are experimental or operational, or both. Program details
are available from the SpectroSystem Manual [21]. However, some information on the AID routine, program modifications, and the algo-
rithms of two routines is presented in the following order.

AID: The computer is programmed in such a way that under the
"WAITING" mode, the "NO" key transfers the control to the "AID"
mode. All inputs in "AID" mode are terminated by a CR. A location
is accessible by entering a "CXXXXX" where XXXXX is the location
address. The computer prints out the content in the location and
waits for further instruction. The desired change can be typed in
or a period (.) is entered to keep the content in the memory. A
comma (,) causes the computer to print out the content of the next
location. A command "VXXXXX." directs the computer to output all
of "SXXXXX,YYYYY,0,0." results in printing of the memory contents
from the location address XXXXX to YYYYY. Some AID operation can be
interrupted and terminated by pushing the "RUB OUT" key on the teletype. A "G3." transmits the computer control back to the "WAITING"
mode.

Modifications: The Varian software package apparently was de-
vised primarily for esr spectroscopic measurements. This type of

experiment usually does not involve the time variable as it does in
kinetic measurements. After attempting several kinetic experiments
using the original Varian software, we found that the time derived
from the computer clock was significantly different from the time
recorded by a stop watch for measurements that took several hours
to complete. We prepared a subroutine to calibrate this time deviation. The program was stored on a paper tape that could be read
into the computer through the tape reader after the Varian system
tape was entered.

The technique of multiple scan was not employed in our experiments with radicals produced by gamma irradiation of PE. Hence,
the MSAV program was not useful to our operation. On the other
hand, we needed a subroutine to print out kinetic data in a format
convenient to the subsequent data processing. To satisfy this need,
we have developed a data output (∅TPT) routine which replaces part
of the MSAV program. The program performs a minor data reduction
before it prints out the results on the teletype in four digits of
decimal form. Because this modified program for kinetic measurements requires use of some of the locations in the SPNC program,
only the kinetic or spin concentration measurement program can exist
in the memory. However, we have prepared a short paper tape that
recovers the SPNC program for spin concentration experiments. This
tape saves us from reloading the entire system tape which takes
approximately 40 minutes.

SMTH: This program performs curve smoothing on the spectrum
data using a quadratic formula [21] and stores the smoothed data
points in the locations for the processed spectrum. The philosophy
of the smoothing is to reduce instrumental noises from the raw data
by applying a statistical weight function.

SPNC: The SPNC program is employed to calculate the spin concentration of an esr spectrum stored in the computer data locations.
The method is equivalent to measuring the area under an absorption
curve, although the area is not determined from the original curve
directly. It is common in esr spectroscopic researches that the

tails of an absorption curve approach the spectrum baseline very slowly. As a result, evaluation of the area under the absorption spectrum becomes difficult. To ignore the tail portions of the spectrum can introduce significant errors, particularly when the baseline is tilted. Generally, this problem can be reduced, if not totally solved, if the area is evaluated by double integration of the first-derivative spectrum. The first-derivative curve can be obtained readily by an analog derivative computer. This is exactly the technique employed in our esr system. Very often, the slope of the absorption curve approaches zero much faster than the tails of the curve approach the baseline. Thus, the trouble encountered in determining the baseline can be reduced. Also, for a particular measurement, the baseline shift or tilt is a constant that can be eliminated by using the derivatives of the original esr absorption curve.

The program uses a formula that assumes the spectrum being investigated is symmetrical [22] about a center. The spectrum "center" point is located using an iterative procedure. The iteration starts by dividing the collected spectrum into two equal halves. The area under each half spectrum curve is evaluated separately, and the two values are compared. The center of the "half" curve which gives the largest area is taken as the dividing point for the next computation. The procedure is repeated until the two areas on each side of the dividing point agree within a specified limit or up to 10 iterations. The area under each "half" spectrum is printed, and the sum of the two values of the last iteration is taken as the total area under the absorption curve.

In theory, this technique of "half" dividing makes possible an approach to the true spectrum center or symmetry point very rapidly. In practice, however, the iteration result can overshoot the criterion point, which is the key to turning off the iteration. As a result, the areas of the two "half" spectra of the last iteration may not be as close to each other as those of a previous iteration. An example is given later in this section. It must be noted that the errors associated with an asymmetric spectrum are not known

for such a calculational method. Fortunately, the spectra of our
present interest are fairly symmetrical, and the deviations derived
from the SPNC program evidently have no significant effects on the
experimental results.

3. Operation

Application of the SS-100 console has simplified the entire operation into simple actions of pushing a button. Some programs require parameters that generally proceed through question-answer
manipulations performed on the console panel and keyboard as shown
in Figs. 13 through 15. Details of the system operation are described in the manual for the SS-100 console [21]. Brief discussions are presented on operating the console and applications of
the SMTH, SPNC, and MSAV programs. Use of other routines are
straightforward.

a. *Console.* The SS-100 console operation is strictly a keyboard maneuver. When a key on the control board (Fig. 13) is depressed or a message is delivered from the computer to the message
board (Fig. 15), the associated signal light underneath the panel
lights up. At completion of the operation, the light automatically
turns off.

The 620/i minicomputer is equipped with an automatic start
that places the computer at "WAITING" mode as soon as the computer
is powered on and the "COMPUTER CONTROL" key is depressed. ("WAIT-
ING" in this case is equivalent to the "SYSTEM?" of the H112 system
package.) To select an operation, the "SELECT OPERATION" key is
depressed and the "MARKER" light is moved downward or upward to the
right of the desired program (Fig. 14), and the "SELECT" key is
pushed. Any desired operation can be deleted by using the "DELETE"
key. The operator then depresses the "ENTER PARAMETER" key to
activate the numerical keynotes. The computer responds with a signal light at the appropriate position of the parameter board. The
operator moves the "PARAMETER MARKER" light next to the indicated
parameter and enters the necessary numbers through the keyboard.
The numbers are displayed on the digital display. A wrong entry

can be deleted by pushing the "CLEAR" key. The numbers can be
loaded into an assigned location by depressing the "LOAD" key. The
"REPEAT" key is used to keep the parameter already in the memory.
Numbers greater than four digits have to be entered through the
teletype in octal form.

The operator has the option of verifying the parameter he has
just entered by depressing the "VERIFY" key and then the "YES".
The program prints out the parameter on the teletype. However,
only numbers of four digits or less can be correctly printed. Apparently, there are "bugs" in the print routine, although they do
not affect our measurement. An operation can be interrupted any
moment by the "IMMEDIATE HALT" key and the control is transferred
back to "WAITING". The "RESUME" key resumes the operation. Obviously, the "IMMEDIATE HALT" in this case is similar to the "TC"
subroutine in the H112 system, but the latter does not have the
option of resuming the operation.

Both the spin concentration measurement and kinetic experiment
are accomplished by selecting the SSCN followed by the options of
data reductions. The operation modes of "SELECT OPERATION," "ENTER
PARAMETER," and "START" must be selected one after the other, i.e.,
the "ENTER PARAMETER" key is depressed before the "START" even if
no parameter is needed in the program. The "START" key initiates
the message of "SCOPE FIELD SW OFF?" which makes sure that the
field switch of the oscilloscope on the spectrometer is off. A
"YES" activates another message to confirm "PEN OK?" which is for
tracing a plot during the measurement. Another "YES" then actually
starts the data acquisition. In all, 1000 data points are collected
in the computer memory, starting from location address 12000. At
completion of the operation, the computer returns to "WAITING." Two
examples of running the spin concentration measurement and kinetic
experiment are given in Figs. 17 and 18. Details of the measurement procedure have been described elsewhere [23-25].

b. SMTH. The number of points for smoothing is the value required in the quadratic formula. In general, 10 to 20 points are

appropriate. Figure 16 shows the result of an allyl radical spectrum before and after the smoothing reduction. The operational
procedure of this particular measurement is given in Fig. 17. The
smoothed spectrum is plotted after multiplying the original spectrum by a selected constant to avoid overlapping with the original
unsmoothed spectrum.

   c. SPNC. The spectrum of Fig. 16 is calculated using the SPNC
program as shown in Fig. 17. The standard concentration (STD.
CØNC.) is the spin concentration of a standard sample used in the
measurement. A zero is entered if no standard is used. The moment

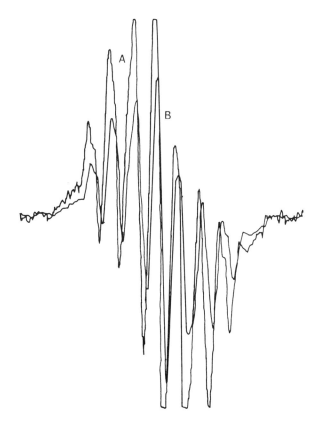

FIG. 16. An allyl radical spectrum measured at 115K before
(A) and after (B) the smoothing.

```
SSCN
  TIME    0360
PLDS
SMTH
  #PTS    0010
PLPS
  PLAM    0080
SPNC

STD.CONC. X.XXEXX
0

MOM'T NUMB X.XXEXX
1

MAG.FLD. IN GAUSS=
4.0E02

% CNTR FLD FOR CALCN X.XXEXX
7.0E01

BAS/LIN CORR. AT ? LOC. X.XXEXX
2.0E01

+1.2934E+07
+0.8275E+07
-0.7146E+06
+1.0901E+07
+0.6165E+06
+1.5003E+07
+0.4903E+07
+1.3448E+07
+0.8413E+07
+1.1259E+07
+1.0080E+07
+1.0205E+07
+1.0997E+07
+0.9621E+07
+1.0617E+07
+0.9869E+07
+1.0437E+07
+0.9986E+07
STD.CONC.=0
NEW CONC.=0
TOTAL MOMENT= +2.0424E+07
```

FIG. 17. Allyl radical moment (area) of the spectrum shown in Fig. 16 as calculated by the SPNC program.

number (MØM'T NUMB) is always 1 in our measurements, although the
program allows for computation of second moments. The definitions
of spin moments can be found in Ref. 22. The magnetic field (MAG.
FLD.) is the field strength of the measurement. Actually, the
field is not used in the determination of the absorption area or
first moment. Nevertheless, care must be taken to consider the
effect of the magnetic field on the final evaluation if the sample
is measured at a MAG. FLD. different from the standard material.
The percent center field for calculation (% CNTR FLD) is the percentage from the center of the spectrum that is used in the computation. The baseline correction (BAS/LN CØRR.) is the location at
which the baseline correction begins. It must be larger than or
equal to 20 points to protect against a possible spurious noise
level at the early portions of the spectrum.

Results of the example shown in Fig. 17 show that the sixth
iteration gives two "half" areas closer to each other than the
values from the ninth iteration, and yet the latter values are used.
However, the results of these two iterations agree to better than
0.7 percent.

d. *MSAV*. Figure 18 shows the operational procedure and results of a typical kinetic run of the allyl radical decay measurement. Data interpretations for this system will be presented in
Sec. III.C of this chapter. For this particular measurement, the
scan time is 24,000 sec, but only the four least significant digits
are printed on "VERIFY." The raw data are smoothed by averaging
each point over 20 points, 10 before and 10 after. The smoothed
data points are stored in locations beginning at 14000. Finally,
the MSAV (OTPT) is selected to output the kinetic data. The program requires the point interval (PTIN), which is the number of
points to skip before outputting one point, to be specified. For
convenience of data processing, usually not more than 50 points are
used. In this particular example, PTIN is 20, which is equal to
8 min per point.

The "START AT" and 14000 direct the computer to use the

```
SSCN
  TIME    4000
PLSS
  PLAM    0080
SMTH
  #PTS    0020
PLPS
  PLAM    0070
OTPT
  PTIN    0020
TIME INT =  8.0E00
START AT  14000
WGHT =  2.50E-02

DOSE =  4.00E01

REC. GAIN  =  5.0E03
```

TIME(MIN)	PK. HGHT.	ATG
+0.8000E+01	7159	+1.4317E+00
+1.6000E+01	6930	+1.3860E+00
+0.2400E+02	6755	+1.3510E+00
+0.3200E+02	6648	+1.3296E+00

DATA OMITTED

```
014000
014000 016432 016457 016416 016402 016302 016341 016233 016236
014010 016260 016262 016204 016166 016150 016130 016113 016073
014020 016055 016031 016006 015767 015752 015735 015715 015676
014030 015660 015643 015630 015611 015575 015566 015556 015543
014040 015525 015511 015476 015466 015453 015440 015427 0154
012000.
012000 016432 016457 016416 016402 016302 016341 016233 016236
012010 016260 016262 016203 016154 016163 016106 016104 016151
012020 016067 015767 016047 016011 015733 015663 015714 015730
012030 015651 015640 015622 015632 015637 015554 015570 015510
012040 015515 015544 015466 015427 015464 015550 015411 015407

03.
```

FIG. 18.   Kinetic data of a typical allyl radical decay experiment measured at 70°C.

smoothed data for the calculation. The program normalizes the measured peak height (PK. HGHT.) to peak height per Mrad, per gram of sample, and per receiver gain (RG). The Mrad is the dosage in megarads
of gamma irradiation that the sample has received. The RG is the
E-4 spectrometer setting. Only part of the 50 points are given in
Fig. 18. The data blocks in octal notation shown at the bottom of
the figure are the smoothed and unsmoothed data points, respectively,
during the early parts of the reaction. They are listed for checking
the "zero-time" value of the run in case it is needed in the calculation.

4. Application

This Varian computer system is mainly for the E-4 spectrometer. It
cannot be employed to interface with other equipment without some
major hardware modifications. It is a very efficient system, although it is not as flexible as the previously described H112 machine.

II. COMPUTER TECHNIQUES IN DATA INTERPRETATION

During the early years of the computer era, the applications of
high-speed electronic computers were primarily in repeated computations of simple problems and solutions of complex systems beyond the
reach of human abilities. In the past decade, computers were promi-
nently employed to the opening of a new area of science known as
Simulation [26]. In rather loose terms, Simulation is the act of
representing some aspects of the real world by numbers or symbols.
To some extent, it is a method of simulating a real experiment using
numerical iterations based on a specified algorithm. Simulation
problems are, practically, unsolvable without the use of high-speed
electronic computers. Recently, computer uses have been broadly extended into various areas of polymer science [27]. Common to all of
these applications is the ability of electronic computers to perform
the tasks of complex computations.

Along with the role of minicomputers in laboratory automation
is their use in performing calculations. With a computer-controlled
instrument, an experimentalist can obtain a much larger volume of

data in a single measurement than he can in several days, or even
months, of work. Under such a circumstance, the need for a computer for computation purposes is even more critical. In fact, a
computer-controlled experiment is never effective without a close
alliance with data processing techniques.

An efficient programing language is required for effective use
of an electronic computer in solving problems of science, engineering, and business. Because the data processing, or interpretation,
is generally performed on a larger computer, high-level languages
are usually employed. The programing techniques with a high-level
language are much easier than the assembly languages or machine
codes such as those employed in the Honeywell H112 and Varian 620/i.
A programmer who knows nothing about assembly language can program
a computer readily with a high-level language of his preference,
presuming that the computer is equipped with a compiler of that particular language. In general, a high-level language is easy to
learn; it has simple "grammar" and "sentence" structures much easier
to comprehend than those in English. The high-level languages of
most common uses are, as previously noted, BASIC, FORTRAN, ALGOL,
COBOL, and PL/I. BASIC, [28] FORTRAN, [28-30] and ALGOL [30,31]
are scientifically oriented. COBOL [28,32] specializes in business
applications. PL/I [16,30,33] is a unique language for both scientific and business applications or the combination of the two.
However, this does not mean that a problem of a business nature
cannot be solved by a scientific language or vice versa. Actually,
some programmers prefer FORTRAN to COBOL in handling some particular
bookkeeping routines.

The learning and understanding of a high-level language involve
only elementary knowledge of mathematics. After one language is
mastered, others are easily acquired because basic programing techniques are common to all languages even though there may be differences in coding. In the following section, fundamental FORTRAN
techniques are described briefly. A sample program is given to
illustrate the uses of various important statements of this language.

Details of the language can be found by reference to the literature
[28-30].

A. Fundamentals of FORTRAN Programming Techniques

FORTRAN or "Formula Translation" is particularly efficient in
handling formula computations common to science and engineering.
There are more than a dozen versions of FORTRAN that differ by minor
variations in details and fundamental alterations in scope. The
FORTRAN techniques of interest to the authors have been applied
equally well on the Honeywell 1620, IBM 360, and CDC 6600 computers,
and the HCN timesharing systems.

Programing is an art that needs more practice than theory. One
often learns more from making mistakes. Also, it is possible to
learn by studying programs written by someone else. Although the
availability of a computer often speeds up the learning procedure,
preliminary studies can be conducted without actually working with a
computer. A basic concept that needs to be emphasized is that a
computer accepts instructions or statements only one at a time and
that it operates on a base of "first come first serve." Therefore,
a thumb logic for programing is advancement of step-by-step. The
computer "forgets" the history of any operation and "remembers" only
the variables that have been assigned or defined.

1. Programing Essentials

As a general rule, the first step of programing is to write a program outline. It is always a good habit to prepare this outline in
form of a flow-diagram similar to those shown in Figs. 7 and 8. In
any computation, simple or complex, there must be operations of data
input, arithmetic, and results output. Obviously, it is of prime
importance to be familiar with the codings of these operations that
differ from one machine to another. Fortunately, the symbols for
arithmetic operations are identical to all languages. In addition
to the conventional data input/output devices, i.e., cards, paper
tapes, magnetic tapes, and magnetic disks, recent investigations
indicate that holographic techniques can be applied to tape memory
systems [34].

In practice, the FORTRAN language can be handled at ease if the statements for looping and transfer operations are mastered. The use of subprogram techniques certainly is very helpful or even necessary for tackling complex problems.

2. Example

The example given in Fig. 19 demonstrates the use of all six major types of FORTRAN statements, i.e., FØRMAT, READ/WRITE, DIMENSIØN, DØ, GØTØ/IF, and CALL that ought to be familiar to the reader. The program calculates the parameters of a line that fits a set of X-Y data pairs using a linear least-squares method [35].

LIST LLSQ

```
 10 C FØRTRAN IV PRØGRAM FØR LINEAR LEAST-SQUARES CALCULATION
 20 C   FOR USE ØN PRØGRAMS SUBMITTED TØ A COMPUTER CENTER
 30      CØMMØN X(100)
 90      WRITE(3,1)
100    1 FØRMAT(//'NØ. PT.',8X,'X',14X,'Y')
110      I=1
120    2 READ(1,3)X(I),Y(I)
130    3 FØRMAT(2F10.0)
140      IF(X(I).LE.0.)GØTØ5
150      WRITE(3,4)I,X(I),Y(I)
160    4 FØRMAT(I5,6X,2(4X,E11.4))
170      I=I+1
180      GØTØ2
190    5 I=I-1
200      CALL LESQ(X,Y,I,A1,A2,SA1,SA2)
210      WRITE(3,6)A1,SA1,A2,SA2
220    6 FØRMAT(//'SLØPE      =',E11.4,5X,'DEV =',E11.4/
230      +'INTERCEPT   =',E11.4,5X,'DEV =',E11.4///)
240      STØP
250      END
260      SUBRØUTINE LESQ(X,Y,N,A1,A2,SA1,SA2)
270 C
280 C   THIS PRØGRAM FITS Y AS A LINEAR FUNCTIØN ØF X
290 C     N=NUMBER ØF X,Y PAIRS
300 C     A1= SLØPE,SA1=STANDARD ERRØR ØF A1
310 C     A2=INTERCEPT, SA2=STANDARD ERRØR ØF A2
320 C
330 C     METHØD ØF LEAST SQUARES, Y. BEERS 'INTRØDUCTIØN TO
```

FIG. 19. A FORTRAN IV program of linear least-squares calculation for timesharing use.

(continued)

In this program, the main program first prints a title of the input parameters, then reads in a data card that contains a pair of X-Y data points. The program prints the data before it reads the next data card. A loop is set up in the program to count the number of cards entered. A blank card can be used to terminate the input and start the computation. The reader may find it very convenient to use this looping technique to enter a series of uncounted data sets.

Actually, for such a simple calculation, the "CALL" instruction on line 200 can be replaced by the lines 380 through 600 to eliminate the subroutine process. Obviously, the variables I and N in

```
340 C     THEØRY OF ERRØR', 2ND ED., ADDISØN-WESLEY PUBLISHING
350 C     CØ., 1962, P. 38
360 C
370       DIMENSIØN X(1),Y(1)
380       SUMX=0.
390       SUMY=0.
400       SUMXY=0.
410       SUMX2=0.
420       SUMY2=0.
430       SMDY2=0.
440       DØ 1 I=1,N
450       SUMN=SUMX+X(I)
460       SUMY=SUMY+Y(I)
470       SUMXY=SUMXY+Y(I)*Y(I)
480       SUMX2=SUMX2+X(I)*X(I)
490     1 SUMY2=SUMY2+Y(I)*Y(I)
500       DEN=N*SUMX2-SUMX*SUMX
510       DA1=N*SUMXY-SUMX*SUMY
520       DA2=SUMX2*SUMX2*SUMY-SUMX*SUMXY
530       A1=DA1/DEN
540       A2=DA2/DEN
550       DØ 2 I=1,N
560       DY=Y(I)-A1*X(I)-A2
570     2 SMDY2=SMDY2+DY*DY
580       SY=SQRT(SMDY2/(N-2))
590       SA1=SY*SQRT(N/DEN)
600       SA2=SY*SQRT(SUMX2/DEN)
610       RETURN
620       END
```

FIG. 19 (continued)

the subroutine selection have to be interchanged and the statement
numbers on lines 440, 490, 550, and 570 must be reassigned so that
no conflict can occur. However, the example is employed primarily
to demonstrate the use of several major FORTRAN statements.

3. Timesharing

It can be very time-consuming if the data reduction process is performed completely by handling of computer cards. It is particularly
tedious with computer-controlled experiments for which data are
punched on paper tapes or recorded on magnetic tapes or disks.
Timesharing techniques make data reduction simple, and results of
calculations can be obtained as soon as the data are read into the
system. The speed of data conversion is always a main concern of
most experimentalists.

There are many systems of timesharing. The PDP-10 and HCN
systems are of interest to the authors. Again, the basic philosophy
of timesharing is identical to all systems although details in programing codes may be different with various machines.

To illustrate the advantage of using the timesharing techniques,
the program in Fig. 19 was modified so that it can be processed on
the HCN system. This modified program is shown in Fig. 20. By careful comparison of the two programs, the reader easily discovers that
the changes are minor and are mainly on the input/output statements.
Lines 40 to 80 have been added in the timesharing program to accommodate the data input procedure. The FORTRAN compiler in the HCN
system has practically the same features as a general FORTRAN compiler. In the example given in Fig. 20, a data file must be created
in the memory so that the program can locate the file for calculation. The output code for the timesharing program has been switched
from 3 to 9, the code for teletype input/output in the HCN system.

Extensive use of timesharing techniques requires further knowledge of operational techniques. The reader is referred to the
timesharing handbooks for the PDP-10 [36] and HCN [37] systems.

B. Thermal Analysis of Polymers

In this section, we discuss some general data interpretation techniques for thermal analysis of polymers in the area of decomposition, curing kinetics of polymers, and compatibility tests of polymers with energetic materials. Measurements were performed on the Honeywell H112 computer-controlled TGA and DSC systems as described in Sec. II. All computer programs in the studies were executed on the HCN timesharing system.

Because our purpose is to demonstrate the applicability of the data analysis techniques to the associated problems, only one example is given on each topic. Also, no attempt is made to study the mechanism of any of the reactions. The data of the curves shown in Figs. 10 and 11 are used as examples of the analyses for TGA and DSC techniques, respectively. The compatibility test methods are exemplified by a system containing Teflon TFE and a propellant.

1. Thermal Decomposition Kinetics of Polymers

During the past decade, many efforts have been dedicated to kinetic analyses of TGA and DSC measurements. The topics have been extensively reviewed in a series of articles [38] and a book [39].

Although TGA measures the weight change and DSC monitors the exothermicity or endothermicity of a substance during a heating process, both analytic methods follow a specific reaction parameter proportional to the amount of the material presented. Therefore, kinetic analyses for both types of experiments are almost identical in many cases.

The basic mathematical models for the nonisothermal methods in TGA and DSC fall into one of five categories: (1) Integral Method, (2) Differential Method, (3) Differential-differential Method, (4) Initial Rate Method, and (5) Nonlinear or Cycle Heating Method [40]. A brief review [34] of these analytic techniques indicates that the Differential Method appears to be superior to the other methods because of its simplicity and accuracy.

The rate expression of a simple decomposition reaction in solid state takes the general form of

```
10 *      FØRTRAN IV PRØGRAM FØR LINEAR-LEAST SQUARES CALCULATIØN
20 *         FØR USE ØN TIMESHARING TERMINALS ØNLY
30        CØMMØN X(100), NAM(6)
40        WRITE(9,1)
50      1 FØRMAT(/'NAME ØF INPUT DATA FILE :')
60        READ(9,10)NAM
70     10 FØRMAT(6A2)
80        CALL DEFINE(1,NAM)
90        WRITE(9,15)
100    15 FØRMAT(//'NØ. PT.',8X,'X',14X,'Y')
110       I=1
120    20 READ(1,30)X(I),Y(I)
130    30 FØRMAT(2F10.0)
140       IF(X(I).LE.0.)GØTØ50
150       WRITE(9,40)I,X(I),Y(I)
160    40 FØRMAT(15,6X,2(4X,E11.4))
170       I=I+1
180       GØTØ20
190    50 I=I-1
200       CALL LESQ(X,Y,I,A1,A2,SA1,SA2)
210       WRITE(9,60)A1,SA1,A2,SA2
220    60 FØRMAT(//'SLØPE       =',E11.4,5X,'DEV ='E11.4/
230       +'INTERCEPT   =',E11.4,5X,'DEV =',E11.4///)
240       STØP
250       END
260       SUBRØUTINE LESQ(X,Y,N,A1,A2,SA1,SA2)
270 *
280 *      THIS PRØGRAM FITS Y AS A LINEAR FUNCTIØN OF X
290 *        N=NUMBER ØF X,Y PAIRS
300 *        A1= SLØPE,SA1=STANDARD ERRØR ØF A1
310 *        A2=INTERCEPT, SA2=STANDARD ERRØR ØF A2
320 *
```

FIG. 20. A FORTRAN IV program of linear least-squares calculation for timesharing use.

```
330 *      METHØD OF LEAST SQUARES, Y. BEERS 'INTRODUCTIØN TØ
340 *      THEØRY OF ERRØR', 2ND., ADDISØN-WESLEY PUBLISHING
350 *      CØ., 1962, P. 38
360 *
370 *      DIMENSIØN X(1),Y(1)
380        SUMX=0.
390        SUMY=0.
400        SUMXY=0.
410        SUMX2=0.
420        SUMY2=0.
430        SMDY2=0.
440        DØ 1 I=I,N
450        SUMX=SUMX+X(I)
460        SUMY=SUMY+Y(I)
470        SUMXY=SUMXY+X(I)*Y(I)
480        SUMX2=SUMX2+X(I)*X(I)
490      1 SUMY2=SUMY2+Y(I)*Y(I)
500        DEN=N*SUMX2-SUMX*SUMX
510        DA1=N*SUMXY-SUMX*SUMY
520        DA2=SUMX2+SUMY-SUMX*SUMXY
530        A1=DA1/DEN
540        A2=DA2/DEN
550        DØ 2 I=1,N
560        DY=Y(I)-A1*X(I)-A2
570      2 SMDY2=SMDY2+DY*DY
580        SY=SQRT(SMDY2/(N-2))
590        SA1=SY*SQRT(N/DEN)
600        SA2=SY*SQRT(SUMX2/DEN)
610        RETURN
620        END
```

FIG. 20 (continued)

$$V = -\frac{d\alpha}{dt} = kW_o^{n-1}f \tag{1}$$

where V and α denote the instantaneous reaction rate and weight
fraction of the sample remaining at time t, respectively; k and W
indicate the rate constant and reaction order, and f is a function
of α and n. For a first-order reaction, f is equal to 1. It can
be readily shown that even for the simplest first-order reaction,
the rate expression, Eq. (1), cannot be integrated in a closed
form for temperature-programmed measurements. Attempts at integrating the equation have been made using series expansion and approximation methods [39,41-43], graphical techniques [44], and the combination of the two [45]. Although the graphical integration method
[44] has been shown to be a useful technique for analysis of TGA
kinetics, it involves tedious graphical methods and can be very
time-consuming. On the other hand, the Differential Method, with
the assistance of computer techniques, has been shown to be a powerful tool for kinetic analysis [46]. For this reason, only details
of the Differential Method are discussed here.

In an experiment, Willard et al. [43] performed a series of
isothermal DSC measurements on the thermal decomposition of an
organic peroxide by a stepwise temperature-jump method at 10°C/min.
They found that the results of this temperature-jump experiment
agreed very well with a measurement at a linear heating rate of
10°C/min. These experimental results of Willard et al. indicate
the applicability of the Arrhenius equation to the nonisothermal
experiments such as the temperature-programed TGA and DSC. By combining the Arrhenius relation with Eq. (1), the derived equation
can be rearranged to

$$\ln\frac{V}{f} = \frac{E_a}{RT} + \ln(AW_o^{n-1}) \tag{2}$$

where T is the temperature in K, R is the gas constant; A and E_a are
the pre-exponential factor and activation energy of the reaction,
respectively. For a temperature-programed experiment with a linear

heating rate r, the rate V can be easily obtained from Eq. (3).

$$V = \frac{d\alpha}{dt} = -r\ \frac{\partial\alpha}{\partial T} \tag{3}$$

where $\partial\alpha/\partial T$ is the partial derivative of the associated quantities. Equations (2) and (3) are the bases of our computer programs for TGA and DSC kinetic analyses. Similar expressions have been derived for TGA [42,45,47,48] and DSC [43,49-52] kinetics.

　　a. Program. A computer program was written to handle all computations for both the TGA and DSC analyses. Reaction rates were obtained from a TGA or DSC curve by a subroutine for numerical differentiation [46] using a second-order Lagrangian interpolation polynomial. The subroutine shown in Figs. 19 and 20 was employed to perform a linear least-squares [35] calculation on Eq. (2), and the kinetic parameters were derived from the slope and intercept of the straight line. Because formats of the TGA and DSC data were slightly different, separated subroutines were prepared for the associated data reductions before they could be used in the general forms of Eqs. (2) and (3).

　　Although the reaction order of the program is a parameter that can be any positive number, the present program applies strictly to first-order kinetics. However, it can be modified easily to handle simple reaction kinetics other than first-order processes. The program can process as many as 750 data points. It also includes a plot subroutine. A user has the option of "plotting" the V versus T and/or ln V versus 1/T of Eq. (2) on the teletype.

　　Prior to execution of the program, a data file must be created in the computer memory by entering a paper tape that contains the data obtained from the H112 system. All the other operations are straightforward and are accomplished through question-answer maneuvers.

　　b. TGA Decomposition Kinetics of TEFLON. The thermal decomposition kinetic data of TEFLON TFE, shown in Fig. 10, were attributed to a first-order reaction. A list of the calculational procedure, including the data and results for this particular run, appears in Fig. 21, and the computer plots of the results are given in Fig. 22. It is clear that the data exhibit a reasonable linearity to the first-order plot (Fig. 22b) up to 98 percent of the reaction.

NAME OF INPUT DATA FILE :! TEFL

ALL NUMERICAL INPUTS ARE IN FORMAT F10.0

DIFFERENTIAL TECHNIQUES FOR! TGA

   THE KINETIC ORDER (N) IS! 1.

PROGRAM HEATING RATE, DEG K/MIN   =° 5.

STARTING TEMPERATURE, DEG K   =! 642.

TIME INTERVAL FOR EACH POINT, (SEC) =! 12.

DO YOU WANT A LIST OUTPUT ? (YES/NO)! YES

T(DEG K)	1/T	FRACN RM	RATE(/SEC)	LN(RATE/(W**N))
0.6420E 03	0.1558E-02	0.1000E 01		
0.7060E 03	0.1416E-02	0.9739E 00	0.8783E-04	-0.9319E 01
0.7070E 03	0.1414E-02	0.9778E 00	0.1775E-03	-0.8614E 01

DATA OMITTED

| 0.7820E 03 | 0.1279E-02 | 0.3409E-01 | 0.9098E-03 | -0.3523E 01 |
| 0.7330E 03 | 0.1277E-02 | 0.2450E-01 | 0.6657E-03 | -0.3605E 01 |

ERROR PARAMETER OF THE NUMERICAL DIFFERENTIATION =   0

NUMBER OF POINTS WITH RATES =/< 0   =   0

SLOPE   =-0.3315E 05   DEV = 0.5296E 03
INTERCEPT = 0.3824E 02   DEV = 0.7123E 00

E(KCAL/MOLE)= 0.6587E 02   DEV = 0.1052E 01
A(PER SEC) = 0.4036E 17

    FIG. 21. Calculational procedure and results of first-order decomposition reaction of TEFLON TFE using the data from Fig. 10.

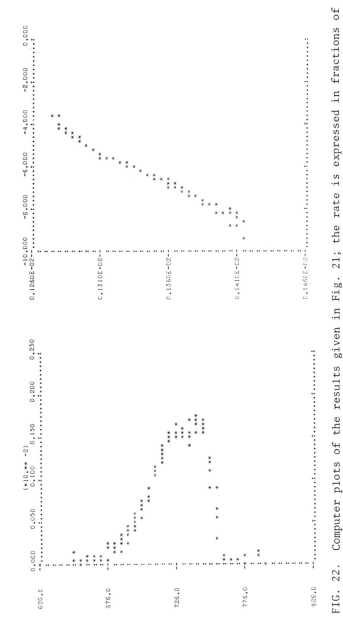

FIG. 22. Computer plots of the results given in Fig. 21; the rate is expressed in fractions of reaction per second.

In Table 1, the kinetic parameters of the present measurement are compared with those from the literature. The activation energy of our measurements appears to be low but compares well with the values of Doyle [41] and Goldfarb, et al. [53], and the pre-exponential factor agrees to within an order of magnitude of that obtained by recalculating the data of Goldfarb et al. [53]. The consistency of the present calculational method also is shown by the same calculation as indicated in Table 1. It appears that the activation energy of the first-order decomposition reaction is higher in vacuo than in nitrogen gas. This is probably associated with the diffusion process of gas products from the bulk sample. However, there is some evidence that bulky powdered samples of TEFLON decompose through a two-stage process of zero-order reaction initially and first-order after 15 to 40 percent of weight loss [57].

TABLE 1

Kinetic Parameters for First-Order Thermal Decomposition of TEFLON
as Determined by the TGA Method

References	Carrier gas	E_a (kcal/mol)	A (sec^{-1})
This work	N_2	65.87 ± 1.05	4.04×10^{16}
Doyle [40]	N_2	67.0	---
Goldfarb et al. [53]	Vacuum	69.34	11.69×10^{16}
Goldfarb et al.[a]	Vacuum	71.98 ± 1.53	0.92×10^{16}
Anderson [54]	Vacuum	75.0 ± 4.0	---
Anderson [48]	Vacuum	77.0 ± 1.0	6.67×10^{18}
Madorsky et al. [55]	Vacuum	80.5	4.67×10^{18}
Siegle and Muus [56]	Vacuum	83.0	4.50×10^{19}

[a]Recalculation from the data of Goldfarb et al., using the present Differential Method.

A mechanism has been proposed [48] to explain the first-order
decomposition kinetics of TEFLON. According to the mechanism, the
decomposition is initiated by homolytic splitting of a C-C bond in
the TEFLON backbone into a free radical pair that stays trapped in
the solid matrices. The radicals move along the chain by stepwise
ejection of monomer molecules, which then diffuse rapidly away from
the reaction sites.

Apparently, the present differential technique is the simplest
calculational method for TGA kinetics of simple reaction order such
as the thermal decomposition kinetics of TEFLON TFE that we have
just described.

2. DSC Curing Kinetics of a Styrene Resin Composite

Although the basic kinetic expressions for calorimetric analysis,
e.g., DSC or Differential Thermal Analysis (DTA), are identical to
those for TGA in many cases, the application of Eq. (2) to DSC
kinetic analysis is not immediately apparent. For a simple decomposition process in which no phase transition is involved, the
amount of the material reacted is proportional to the amount of the
observed exothermicity. Using this, Eq. (2) can be transformed
into a form suitable for calorimetric applications. This expression is given in Eq. (4) for a first-order reaction

$$\ln \frac{rate}{S_t - S} = -\frac{E_a}{RT} + \ln A \qquad (4)$$

where S is the area under the DSC or DTA curve at time t, and S_t is
the total area at completion of the decomposition. The reaction
rate, Eq. (3), in this case is simply the rate of exothermicity
measured by the DSC or DTA. Equation (4) also can be derived readily from the total differential equation for the enthalpy change of
the system and the basic thermal dynamic equations.

Similar to other spectroscopic measurements, the DSC experiment encounters the difficulty of baseline determination, and is
even more severe because of experimental difficulties. The cause
of baseline shift before and after a measurement is inherent to the

thermal properties of the sample and reference cells. In some instances, the shift can be so great that if the baseline cannot be
determined accurately, integrating the curve is almost impossible.
Several methods involving the drawing of baseline for DSC thermo-
grams have been reported [19,58,59]. Most of these methods use
graphical and computer methods.

In the present data interpretation techniques, the area enclosed by a DSC curve and the associated baseline was determined
by numerical integration using a trapezoidal formula. We have
shown that application of the Simpson's rule [60] in the integration does not significantly improve the precision of the integral
result. In the baseline determination, 20 points at the beginning
and another 20 points at the end of the DSC curve were used to determine two separate straight lines using the linear least-squares
techniques discussed previously [35]. The derived 20th point of
the first line and the first point of the second line were employed
to draw the baseline for the DSC curve. This baseline method was
found sufficient for our computer-controlled experiments.

The data of the curing of a styrene resin composite in Fig. 11
were analyzed by the aforementioned interpretation techniques using
the computer program described in the previous section. Formulation of this specific composite is shown in Table 2. This plastic
material has been employed in fabrication of plates and cups for
serving food and drinks on board planes. Because the material involves a proprietary formulation of the source, detailed information of the resin is not known. However, it is clear that the resin
contains styrene monomers and other prepolymers. Results of the
first-order calculations of the data are traced by the computer on
Fig. 23. The Arrhenius plot of Fig. 23b is not linear, and the
curing process is clearly not a first-order reaction with respect
to the total amount of the remaining materials. Because the composite is a highly filled plastic and the curing product is actually
a copolymer of styrene with the resin, it is not at all surprising
that the curing kinetics do not obey a first-order process. In

TABLE 2

A Styrene-resin Composite Formulation

Ingredient	Weight %
Styrene-resin	16.9
Catalyst	0.118
Stearate	0.481
Melamine	0.025
Calcium carbonate	74.6
Fiberglass	7.95

fact, pure styrene polymerizes through a second-order [60] process with respect to styrene.

We observed some first-order decomposition reactions of gun powders using the DSC method, and the associated data interpretation techniques have been successfully applied to these systems. However, the topic of thermal decomposition kinetics of explosives and propellants is obviously not our present interest.

3. Compatibility Testing of Polymers with
   Energetic Materials

During the 20th century, the use of plastic materials in ammunitions have been one of the rapid developing areas in polymer applications. In parallel with the developments, techniques of compatibility testing for polymers with energetic materials become increasingly important. Conventionally, compatibility tests are conducted according to the Vacuum Stability Method [62]. In general, this method involves the use of 5.0 g of an energetic material (explosive or propellant) mixing with 0.5 g of a polymer. The mixture is evacuated and kept at 90°C for 40 hr. The amount of gas evolved at the end of the test minus that obtained from control measurements of the pure components is used as the index of stability of the system. However, the definitions in this test method are ambiguous, and we have obtained evidence that the Vacuum Stability Method fails to provide correct compatibility results for a noncompatible system.

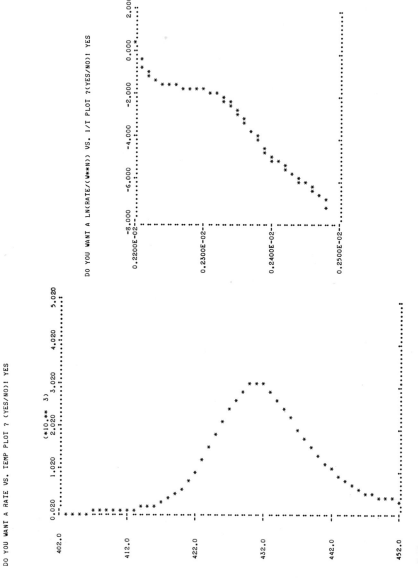

FIG. 23. Computer plots of a styrene-resin composite curing kinetics; the rate is expressed in an arbitrary unit that is proportional to heat evolution.

Accordingly, the scientists at Honeywell have obtained alternative techniques for compatibility test using the thermal analyses, that is, TGA, DSC, and differential thermal analysis (DTA). The TGA technique has been satisfactory for testing, although it is a new technique still under exploration. The DSC and DTA techniques have been documented [63,64].

In general, thermal decomposition kinetics of most explosives or propellants are very complex, and their reaction mechanisms are not known. Because of these difficulties, the approach of compatibility testing through kinetic analysis has been found practically impossible. For this reason, our analytical technique is entirely based on empirical methods.

a. *TGA Compatibility Test of a TEFLON-PYRO Mixture.* In a TGA measurement, the remaining sample weight was monitored by the temperature-programed experiment. Control runs were performed on pure TEFLON TFE and pure PYRO, a nitrocelluloid base propellant. If no reaction occurs between TEFLON and the propellant, the remaining weight fraction of a sample mixture can be calculated from Eq. (5),

$$\alpha = \alpha_e + (1 - \gamma)\alpha_f \qquad (5)$$

where γ is the fraction of the propellant in the starting propellant-polymer mixture; α_e and α_f indicate the fractions remaining in the control runs of pure propellant and pure polymer, respectively. If the TEFLON is compatible with PYRO, the experimental curve of a TEFLON-PYRO mixture should overlap, to a certain extent, or lie underneath that calculated from Eq. (5). Obviously, Eq. (5) is valid only when all measurements are performed with identical program heating rate. A similar test method has been applied to predicting the flammability of composite textile fabrics and polymers [65].

A computer program was developed to perform the necessary calculations of Eq. (5) using the data from two separate control TGA experiments. Results of these calculations, together with the measurements of the mixture, appear in Fig. 24. The deviations of the experimental curve from that calculated from Eq. (5) is less than 0.002% by weight everywhere in the indicated temperature range. At

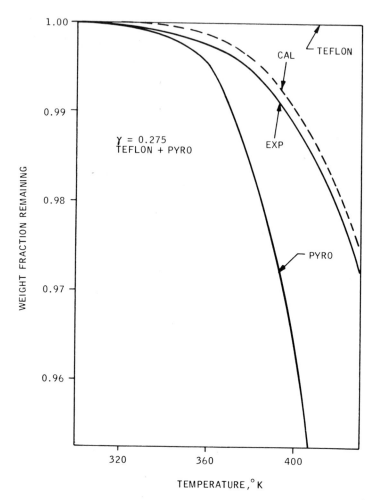

      FIG. 24. Compatibility of TEFLON TFE and PYRO determined by
TGA using Eq. (5); all measurements were performed at a heating rate
of 10°C/min.

approximately 443 K (170°C) PYRO deflagrated or exploded, and the
sample weight decreased extremely rapidly. Based on this experiment, the TEFLON is found to be compatible with PYRO.

      The major difficulty in applying the TGA technique to systems
containing explosives or propellants is that these energetic materials generally have thermal runaway temperatures. As a result,

only less than 10% of the total weight loss during a decomposition process can be accurately determined.

It must be noted that in using the TGA, DSC, and DTA analyses for compatibility tests, a major assumption is imposed that the results of a measurement at high temperatures can be extrapolated to low temperatures; i.e., if two substances react with each other at high temperatures they must also react at low temperatures. The weakness of this definition is that, essentially, any two materials react with each other to some extent and therefore there are no two substances compatible with each other. Apparently, the definition of compatibility is ambiguous and needs clarification. A symposium specializing in the area of compatibility testing [66] has been organized.

C. Electron Spin Resonance Measurements of Irradiated Polyethylene

The use of minicomputers in esr studies or irradiated PE has been most helpful in determining absolute spin concentrations (i.e., number of unpaired electrons per cm^3) in "smoothing" of peak height-time curves, in measuring the rate of alkyl or allyl radical decay, and in subtracting the esr spectrum of a known free radical from that of a mixture in order to obtain the esr spectrum of the other component of the mixture. For recent reviews of this work, see Dole [67,68]. Klopfenstein et al. [6] have recently described their system for the study of esr spectra. Their paper contains a discussion of the baseline correction factor, of their method of subtracting a single component from a complex spectrum, of time averaging and spectral smoothing, of rescaling related experimental spectra for ease of comparison (normalizing first-derivative spectra), and of simulating esr spectra.

1. Determination of Absolute Spin Concentrations and their Variation with Time

Using a Varian E-4 esr spectrometer, a first derivative recording of the esr spectrum is made--or with a Varian 620/i minicomputer interfaced through a SpectroSystem-100 console to the E-4 spectrometer,

the esr spectrum, i.e., the first derivative of the intensity of
spin resonance vs magnetic field, can be stored as mentioned above
up to 1000 data points. Also, with a computer program supplied by
Varian, the first-derivative curve can be doubly integrated and a
number obtained which is proportional to the absolute spin concentration. Another possibility is to punch out the spectrum on paper
tape and store it for future analysis.

The doubly integrated area of the first-derivative spectrum
can be calibrated in terms of known spin concentrations by either
using a freshly prepared solution of diphenylpicrylhydrazyl (DPPH)
in benzene [23] or by mixing uniformly some solid DPPH with powdered KCl [24,25]. The latter is recommended for calibrations at
liquid nitrogen temperature, 77 K. Another possibility is to calibrate a standard pitch sample at room temperature and use this
standard at 77 K. One cannot compare peak heights at different
temperatures because the peak heights vary with temperature [25]
depending on the type of sample.

In kinetic studies--the measurement of the decrease of esr peak
heights with time--it is extremely helpful to have a computer-
controlled automatic smoothing process of the peak height vs time
curves. Figure 16 illustrates a smoothing of the whole allyl radical spectrum measured at 115 K, while in Fig. 25 an actual esr recording of an alkyl radical peak height as a function of time is
compared with a computer smoothed recording of the same data. The
smoothed curve has been displaced from the curve of the actual data
by multiplying the latter by a constant factor [23]. It is reasonable that the smoothing operation reduces the instrumental noise,
and therefore the data read from the smoothed curve should give a
more accurate representation of the free radical decay rates than
the unsmoothed data.

Another very important capability in kinetic studies is the
ability of the esr spectrometer-minicomputer system to take and
store data quickly because in heating an irradiated PE sample to
room temperature or above, the free radicals decay rapidly.

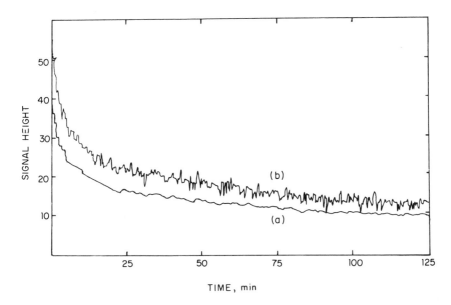

FIG. 25. Comparison of a direct esr recording of alkyl decay
in PE, curve (b), with a computer automatically smoothed curve of
the same data, curve (a).

[Parenthetically, we might add that according to some recent unpub-
lished work of Patel and Dole it has been found possible to produce
alkyl radicals that do not decay at room temperature. These radicals were formed in extended chain samples of PE at 77 K and heated
to 113 K where the evolved hydrogen can be pumped off and the samples returned to room temperature. The residual alkyl radicals in
the vacuum are then stable because there is no hydrogen present to
catalyze their decay]. Obviously, peak heights must be measured,
because it takes considerable time to scan a whole spectrum.

For significant kinetic data to be obtained, it is necessary
to have esr peak purity; that is, the esr spectrum should not be a
mixture of spectra of different radicals, or if it is, a particular
peak should be selected for study that is free of any contributions
from other peaks. In the alkyl radical studies described below,
the outermost or "wing" peak can be used because the allyl radical

spectrum does not overlap with the alkyl spectrum at that point. Conversely, to make accurate studies of allyl decay, all alkyl radicals must first be eliminated by using hydrogen gas to catalyze their decay.

2. Alkyl Radical Decay Kinetics

Johnson et al. [23] found in PE irradiated at 77 K that the alkyl radicals surviving a fairly rapid (5 min) heating to room temperature decay according to two simultaneous first-order decay reactions. By combining two first-order equations it is easy to show that

$$\ln \frac{C-C_s}{C_s} = \ln \frac{C_f^{\,\circ}}{C_s^{\,\circ}} - (k_f - k_s)t \tag{6}$$

where C is the total alkyl radical concentration at time t, C_s is the concentration of the slowly decaying radicals at time t, $C_f^{\,\circ}$ and $C_s^{\,\circ}$ are the concentrations of the fast and slowly decaying radicals at zero time, and k_f and k_s are the first-order decay constants of the fast and slow radical decays. At long times, when all the fast decaying radicals have decayed, C_s can be determined and k_s calculated. Knowing k_s, C_s at short times can be estimated and the left-hand-side (l.h.s.) of Eq. (6) $\ln[(C - C_s)/C_s]$, calculated and plotted as a function of t as in Fig. 26. The slope gives $(k_f - k_s)$ and the intercept at zero time, $\ln(C_f^{\,\circ}/C_s^{\,\circ})$. During the acquisition and storage of the data, the minicomputer could be programmed to measure the alkyl radical peak heights at any selected and constant time intervals. For first-order kinetics it is not necessary to measure absolute spin concentrations. The fact that the data for two different radiation doses, that is, two different initial radical concentrations, fall on the same line (Fig. 26) demonstrates the absence of any second-order component in the decay kinetics [23].

Least-squares reduction of the data, calculation of the k_s and k_f constants, etc., were performed at Baylor University on a CDC 6600 terminal from the University of Texas at Austin, as previously mentioned.

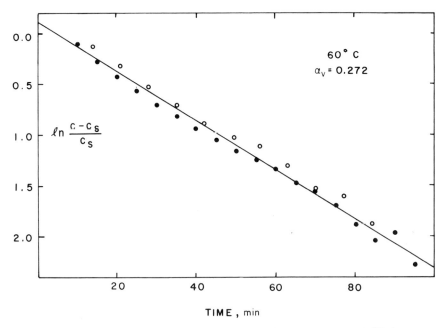

FIG. 26. Test of Eq. (6), alkyl radical decay at 60°C in vacuo: open circles, 14 mrads dose; closed circles, 22 mrads dose; film as-received PE sample.

3. Allyl Radical Decay Kinetics

In PE, the allyl radicals are very stable at room temperature, persisting for months in a vacuum. Gaseous hydrogen has no catalytic effect on their decay. By heating to 60°C or above, the allyl radicals decay by diffusion-controlled, second-order kinetics [25]. The Waite [69] equation for a diffusion-controlled, second-order reaction can be put into the form [25,67] where C and C_o are the

$$(\frac{1}{C} - \frac{1}{C_o})t^{-1/2} = \kappa t^{1/2} + I \tag{7}$$

spin concentrations at time t and zero time, respectively, κ is the slope equal to $4\pi r_o D$, and I the intercept equal to $2r_o \kappa (\pi D)^{1/2}$ of a plot of the l.h.s. of Eq. (7). D is the sum of the diffusion co-efficients of the reacting species, and r_o is the radius of the re-action cage. The reaction cage is defined as the zone within which

the radicals react and outside of which the probability of finding
the radical in any differential volume element is independent of
its position.

Figure 27 illustrates the application of Eq. (7) to the decay
of the allyl free radicals in irradiated single crystals (SC) of PE
[25]. In making these measurements, the minicomputer was programed
to measure peak heights at definite and equal time intervals after
a smoothing process described above. Calculation of the constants
of the Waite equation, Eq. (7), was performed similar to the work
on the alkyl radical decay, on the University of Texas at Austin
CDC 6600 computer.

It should be noted that the initial radical concentration and
the time should be measured as nearly as possible to time zero,
that is, to the inception of the reaction because the radical distribution changes with time as a result of diffusion and reaction.
If one delays too long, the diffusion-controlled nature of the reaction will be undetectable--the intercept I of Eq. (7) will be
zero under which condition Eq. (7) becomes a second-order reaction
without diffusion control. Use of a minicomputer in this work has
materially assisted the taking of quick, uniformly time-spaced allyl
radical decay data.

D. Sol-Gel Evaluations

An extremely important result of the high energy irradiation
of PE is production of interchain carbon to carbon covalent bonds,
or crosslinks. These crosslinks, although few in number, have a
profound effect upon the physical properties of the PE, turning it
into an elastomer above the melting point of the PE, and forming
insoluble three-dimensional crosslinked networks. In fact, radiation crosslinking of PE has led to the growth of a multimillion
dollar business in the United States and around the world.

From a theoretical standpoint it is important to know the
number of crosslinks formed in each gram of sample per unit dose,
but it is very difficult to determine this number. As yet, no

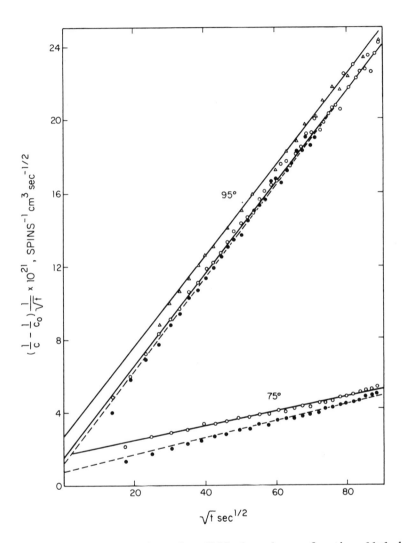

FIG. 27. Test of the Waite diffusion theory for the allyl decay in the single crystal (SC) sample of PE at 75° and 95°C plotted
according to Eq. (7). Dotted lines at 75° for data obtained after
a time delay of 10 min and of 6 min for the 95° data (solid circles).
Solid line and triangles, time advancement of 12 min.

spectroscopic technique has been developed to do this, and crosslinks calculated from swelling or force of retraction measurements
have a rather large uncertainty because of the complicating effect
of mechanical chain entanglements. It is best to estimate the
crosslink yield from dose-gel curves, that is, from data giving the
fraction of insoluble PE produced at a definite dose. The insoluble
fraction is determined by extracting the soluble fraction in boiling
xylene until a constant residual weight is obtained. The irradiated
PE should be annealed in vacuum to eliminate all free radicals before exposure to air.

In the case of PE, the calculation of the crosslink yield is
complicated by the fact that the molecular weight distribution (MWD)
is not the most probable distribution for which a rather simple sol-
gel equation has been developed by Charlesby and Pinner [70]. In
the case of high-density linear PE, MWD more nearly follows the
distribution developed by Wesslau [71]

$$m(P,0) = \frac{1}{\beta\sqrt{\pi}\ P^2} \exp\ [\ \frac{1}{\beta^2}\ (\ln \frac{P}{P(0)}\ e^{-\beta^2/4})^2\ ] \qquad (8)$$

where $m(P,0)$ is the number of polymer molecules at zero dose per
structural unit with degree of polymerization equal to P, and β is
a parameter describing the breadth of the initial distribution and
is equal to $(2 \ln b)^{1/2}$, where b is the initial ratio of weight-
average $(\overline{M}_w)$ to number-average $(\overline{M}_n)$ molecular weight, $\overline{M}_w/\overline{M}_n$. The
mathematical treatment in which the Wesslau distribution has been
incorporated into sol-gel theory is described by Saito et al. [72]
and will not be repeated here. The computer solution of the resulting gel-crosslink equations can be found in the paper by Kang et al.
[73]. In brief, this solution consists first in selecting a number
of Z values where Z is given by Eq. (9).

$$Z = (2g + \lambda)T \qquad (9)$$

in which g is the gel fraction, λ is the ratio of crosslinking and
chain scission G values (i.e., yields per 100 ev of radiation energy
absorbed) and T is defined by Eq. (10).

$$T = y^{-\beta^2/4} \tag{10}$$

in which y is the number of crosslinks per initial number average
molecule. The solution was worked out for nine different β values
ranging from 0.5 to 3.0 and for λ equal to 0. 0.5, 1.0, and 1.5.

In solving the sol-gel equation, two integrals I_1 and I_2 were
calculated by means of a computer program [73] making use of
Simpson's rule [60]. The general flow chart of the computer program is shown in Fig. 28. After the integrals I_1 and I_2 had been
evaluated at each chosen Z value, a value of g was selected, the
function of T calculated from Eq. (9), and right-hand-side (r.h.s.)
and the l.h.s. of the gel-dose (dose assumed proportional to y)

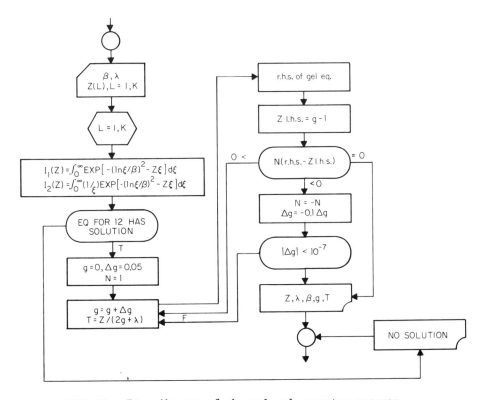

FIG. 28. Flow diagram of the sol-gel computer program.

equation calculated. If the r.h.s. was greater or less than the
l.h.s., g was increased or decreased by a selected amount and the
calculation repeated until the l.h.s. (g-1) was equal to the r.h.s.
within 10^{-7}. The quantity y was then calculated from Eq. (10) and
the functions Z, λ, β, g, T, and y printed out.

In polymer radiation studies it is particularly important to
know the dose to the gel point, i.e., the dose which produces the
first bit of gel (this dose is obtained by plotting g as a function
of the dose and extrapolating to zero g). At the gel point, we
have the two equations, Eqs. (11a) and (11b)

$$y_g = \frac{1 - I(Z_g)}{\lambda(1 - 4\lambda)}$$ (11a)

$$y_g = Z_g \exp(\beta^2/4)/\lambda$$ (11b)

in which

$$I(Z_g) = \frac{I_2(Z_g)}{\beta\pi^{1/2}}$$ (12)

and

$$I_2(Z) = \int_0^\infty \frac{1}{\zeta} \exp[-\ln\zeta/\beta^2 - (2g + \lambda)T\zeta]\, d\zeta$$ (13)

where

$$\zeta = \frac{P}{P_n(0)} e^{-\beta^2/4}$$ (14)

Equations (11a) and (11b) were then solved for y_g by calculating
the r.h.s. of each for various values of Z_g, by plotting y_g as a
function of Z_g for each equation, and by finding the point of intersection of the two curves. Kang et al. [73] performed the latter
calculation with a desk calculator. The other calculations described above could never have been done in this manner. We estimated that it would have taken three thousand years to perform the
calculations with a desk calculator.

IV. CONCLUSIONS

Apparently, computer techniques in both instrumental and computational applications are very broad, and it is almost impossible to predict future developments in either area. Recent developments indicate that computerized instrumentation is taking over many sophisticated analyses, both in industry and pure research. It is foreseeable that laboratory automation will be a booming business in the next decade. The computer has always been the most reliable and efficient "super" calculator in science, engineering, and business; with the rapid growth of computer uses in instrumentation, the demand for a "super" calculator in data reduction is even higher.

Although computer techniques for instrumentation and calculation are apparently different, the basic programing techniques are aimed at interfacing a computer with a particular analysis device. The languages for the task are usually new to most chemists. On the other hand, data reduction can be performed with a wide variety of high level languages usually easy to learn. Generally, a software package for one experimental purpose can be easily adapted to another experiment without major modifications, because most measurements involve a common requirement of measuring the change of one or two reaction variables during the experiment. This is particularly true for dynamic testing in which the desired variables are monitored as a function of time. For this reason, a software package for instrumental controls generally requires no revision if it is well developed. This fact has been shown in the H112 computer control system: two different types of thermal analysis, i.e., TGA and DSC, have been automated by an identical software package. Recently, this control system was successfully coupled with a Cahn accessory for sedimentation analysis and with a free torsion pendulum tester; the results of this study will be reported later. In fact, the same H112 control unit is ready for interface with other equipment for conventional thermal analysis, e.g., DTA, thermomechanical analysis (TMA), Evolved Gas Analysis (EGA), etc. However,

computer programs for data reduction are necessarily different from
one system to another. As far as software is concerned, one can be
confident that a fully developed package can be applicable to many
different types of experiments, but there can be no single computer
program that can interpret two different types of experimental data.

Before ending the chapter, a brief description is given concerning the directions of present and future developments for computer use at the Honeywell Plastics Laboratory.

Recently, experiments were performed by coupling the TGA with
an analog computer to obtain the time derivatives of a weight loss
curve. Preliminary results indicate that the time derivatives obtained by the analog computer have a higher precision than those by
application of the numerical differentiation techniques on the
original TGA curve. It seems that an analog computer can improve
the data accuracy and shorten the data reduction process. Apparently, this derivative technique by an analog computer can also be
applied to other types of measurements.

In the area of material testing for polymers, study is in progress to interface the H112 computer system with an airgun [74] for
high-speed testing. Both tensile and shear strengths of a material
can be measured by this airgun using slightly different test fix-
tures. A long-term project of computerizing other material testing
techniques is also in the plan.

It has been mentioned early in this chapter that the H112 software package is not flexible mainly because a low-level mnemonic
language is required in the programing technique. Computer experts
at Honeywell have developed a FORTRAN compiler, which can compile
FORTRAN statements into DAP-12 (Digital Assembly Programming-12), a
language compatible with SAP-12. It is possible, of course, that a
similar FORTRAN compiler can be prepared for SAP-12 or the mnemonic
language we have been using with the H112 system. When such a compiler exists, anybody who is familiar with the FORTRAN language can
write his own EXEC or main program in FORTRAN for performing the
specific experiment as long as he is aware of all the codes of subroutine already in the software for various tasks. For the present

H112 system, the most convenient operation is to write the EXEC on a timesharing terminal and output the compiled object tape by paper tape punch. This object tape can be entered directly into the H112 for use. Such a method of handling low-level assembly language programing using a high-level language approach [20] has been proved effective and apparently appeals to most chemists who are unfamiliar with assembly languages.

As a conclusion to this chapter, we would like to remind the reader that the computer techniques we have discussed are not limited to use in high polymer chemistry; they are also applicable to other areas of science and engineering.

ACKNOWLEDGMENT

Walter Y. Wen acknowledges C. Olmstead in preparing the H112 System Software, T. Hanwick in reviewing the section on "Minicomputers in Instrumentation," and R. C. Peterson in commenting on the section on "Fundamentals of FORTRAN Programming Techniques." Additional thanks to G. Bellen of Dupont for kindly supplying the TEFLON sample, J. W. Lin for bringing to the author's attention many helpful references, and to Professors C. E. Klopfenstein and O. H. Griffith of the University of Oregon for sending reprints of their papers. The involvement of Walter Y. Wen in this work is made possible by Honeywell Inc.

The research at Baylor University was supported by the U.S. Atomic Energy Commission and income from the Welch Chair in Chemistry endowed by a gift from the Robert A. Welch Foundation.

The authors wish to thank D. R. Johnson for his efforts in modifying the Varian 620/i software package and G. N. Patel for his helpful comments and suggestions.

REFERENCES

1. G. R. Peterson, *Basic Analog Computation*, Macmillan Company, New York, 1967.

2. C. H. Orr and J. A. Norris (ed.), *Computers in Analytical Chemistry*, Plenum Press, New York, 1970.

3. C. E. Klopfenstein and C. L. Wilkins, *Multicomputer Processing in Laboratory Automation*, Fall Joint Computer Conference, 1971, p. 435.

4. C. E. Klopfenstein, *J. Chromatog. Sci.*, *10*, 22 (1972).

5. C. L. Wilkins and C. E. Klopfenstein, *Chem. Tech.*, *Sept.*, 561 (1972); *ibid.*, *Nov.*, 681 (1972).

6. C. E. Klopfenstein, P. Jost, and O. H. Griffith, *Computer Chem. Res.*, *1*, 175 (1972).

7. R. J. Higgins and J. C. Nosler, *Rev. Sci. Instrum.*, *45*, 371 (1974).

8. B. Soucek, *Minicomputers in Data Processing and Simulation*, Macmillan, New Jersey, 1972.

9. J. R. Hileman, *Chem. Eng.*, *May 29*, 61 (1972).

10. S. M. Suchecki, *Textile Ind.*, *March*, 87 (1973).

11. *Minicomputer that Runs the Factory: A Special Report, Bus. Week, Dec. 8*, 68 (1973).

12. *Building Block Approach Offers Inexpensive Process Control, Computer Design, 11; March*, 48 (1972).

13. R. Anderson, *Datamation*, *18*, 46 (1972).

14. D. G. Dykes, *Instr. Control Sys.*, *46*, *March*, 63 (1973).

15. *The Year Mini Users Shed the Security Blanket, Infosystem, 1*, 24 (1974).

16. B. Hodge, *Computers for Engineers: Introduction to Computing Machines and Programming*, Doubleday and Co., Inc., Garden City, New York (1967).

17. T. G. Scott, *Bssic Computer Programming*, Doubleday and Co., Inc., Garden City, New York (1967).

18. G. Romco, E. Lifshin, M. F. Ciccarelli, and D. B. Soresen, *Anal. Chem.*, *45*, 2444 (1973).

19. P. Peyser and W. D. Bascom, Presented at the American Chemical Society Meeting Los Angeles, April 1974; to be published in *Analytical Calorimetry*, Vol. 3, Plenum Press.

20. C. E. Klopfenstein, P. Jost, and O. H. Griffith, *Computer in Chemical and Biochemical Research* (C. E. Klopfenstein and C. L. Wilkins, eds.), Academic Press, 1972.

21. *Operating Instructions, E-4/SS100 with Console*, 87-131-226, Varian Associates, Palo Alto, California.

22. P. B. Ayscough, *Electron Spin Resonance in Chemistry*, Methuen Co., London, 1967, p. 442.

23. D. R. Johnson, W. Y. Wen, and Malcolm Dole, *J. Phys. Chem.*, *77*, 2174 (1974).

24. W. Y. Wen, D. R. Johnson, and Malcolm Dole, *Macromolecules, 7.* 199 (1974).

25. W. Y. Wen, D. R. Johnson, and Malcolm Dole, *J. Phys. Chem., 78,* 1798 (1974).

26. J. McLeod, *Simulation: The Dynamic Modeling of Ideas and Systems with Computers,* McGraw-Hill, New York, 1968.

27. J. B. Kinsinger (ed.), *The Computer in Polymer Science: J. Polymer Sci.,* Part C., Polymer Symposia, New York, 1968.

28. H. Katzan, Jr., *Introduction to Programming Languages,* Macmillan, New York (1973).

29. D. D. McCracken, *A Guide to FORTRAN Programming,* John Wiley and Sons, Inc., New York (1966).

30. C. H. Thomas, *Computers in Chemistry and Instrumentation* (J. S. Mattson, H. B. Mark, and H. C. MacDonald, eds.), Vol. I, Marcel Dekker Inc., 1973, p. 189.

31. D. D. McCracken, *A Guide to ALGOL Programming,* John Wiley and Sons, Inc., New York (1962).

32. D. D. McCracken, *A Guide to COBOL Programming,* John Wiley and Sons, Inc., New York (1970).

33. J. K. Hughes, *PL/I Programming,* Macmillan, New York (1973).

34. J. W. Lin and W. Y. Wen, *J. Thermal Analysis, 9,* 205 (1976).

35. Y. Beers, *Introduction to the Theory of Errors,* (2nd ed.), Addison Wesley Publishing Co., 1962, p. 38.

36. *PDP-10 Timesharing Handbook; PDP-10 Handbook Series,* Digital Equipment Corporation (1970).

37. *Timesharing Documentation, Users Manual; H1640 System,* Honeywell Inc., Doc. No. 70130072330A, Order No. M-1424 (1970).

38. C. B. Murphy, *Anal. Chem., 36,* 247R (1964); ibid., *38,* 443R (1966); ibid., *40,* 380R (1968); ibid., *44,* 513R (1972).

39. L. Reich and S. S. Stivalla, *Elements of Polymer Degradation,* McGraw-Hill Co., New York, 1971, p. 81.

40. C. H. Flynn and L. A. Wall, *J. Res. Natl. Bur. Std., A70,* 487 (1966).

41. C. D. Doyle, *J. Appl. Poly. Sci., 5,* 285 (1961).

42. H. L. Friedman, *J. Poly. Sci., C6,* 183 (1965).

43. P. E. Willard, M. Alvarez, and L. C. Cha, *Poly. Engr. Sci., 11,* 160 (1971).

44. W. Y. Wen, *J. Thermal Analysis, 10,* (1976).

45. V. Šatava and F. Skvára, *J. Am. Ceram. Soc., 52,* 591 (1969).

46. W. Y. Wen, *Intern. J. Chem. Kinet., 5,* 621 (1973).

47. J. H. Sharp and S. A. Wentworth, *Anal. Chem.*, *41*, 2060 (1969).

48. H. C. Anderson, *J. Poly. Sci.*, *C6*, 175 (1963).

49. K. E. J. Barrett, *J. Appl. Poly. Sci.*, *11*, 1617 (1967).

59. G. Beech, *J. Chem. Soc.*, *A, Part II*, 1903 (1968).

51. R. N. Roger and L. C. Smith, *Thermochim. Acta*, *1*, 1 (1970).

52. F. Skvara and V. Satava, *J. Thermal Anal.*, *2*, 325 (1970).

53. I. J. Goldfarb, R. McGuehan, and A. C. Meeks, *Kinetic Analysis of Thermogravimetry; Part II,* Technical Report, AFML-LR-68-181, Part II, Air Force Material Laboratory, Wright-Patterson Air Force Base, Ohio, Sept., 1968.

54. H. C. Anderson, *Makromol. Chem.*, *51*, 233 (1962).

55. S. L. Madorsky, V. E. Hart, S. Straus, and V. A. Sedlark, *J. Res. Natl. Bur. Std.*, *51*, 327 (1953).

56. J. C. Siegle and L. T. Muus, paper presented at 130th American Chemical Society Meeting, Sept., 1956.

57. C. D. Doyle, WADD Technical Report 60-283 (1960).

58. W. P. Brennan, B. Miller, and J. C. Whitwell, *I&EC Fundamentals*, *8*, 314 (1969).

59. C. M. Guttman and J. H. Flynn, *Anal. Chem.*, *45*, 408 (1973).

60. H. Margenau and G. M. Murphy, *The Mathematics of Physics and Chemistry* (2nd ed.), Van Nostrand, N. Y., 1956, p. 477.

61. F. D. W. Billmeyer, Jr., *Text Book of Polymer Science* (2nd ed.), Wiley-Interscience, New York, 1962, p. 313.

62. MIL-STD-286B 1 December 1967, Method 403.1.2 "Vacuum Stability Tests (90 and 100°C)."

63. F. D. Swanson and J. L. Madsen, *Thermal Analysis Testing to Determine the Compatibility of Propellants with Plastics,* Final Report, Contract N00174-72-C-0338, May, 1973, Honeywell Inc., Hopkins, Minnesota.

64. N. E Beach and V. K. Canfield, *Compatibility of Explosives With Polymers (IV): An Addendum to Picatinny Arsenal TR 2595 and PLASTEC Report 33,* PLASTEC Arsenal, Dover, New Jersey, Jan., 1971, p. 76.

65. C. Z. Carroll-Porczynski, *Composites, Jan.,* 9 (1973).

66. F. D. Swanson, Chairman, *Conference on Compatibility of Energetic Materials with Plastics and Additives,* Picatinny Arsenal, Dover, New Jersey, December 3-4, 1974 and Naval Ordnance Station, Indian Head, Maryland, April 27-29, 1975.

67. M. Dole, *Advances in Radiation Chemistry* (M. Burton and J. L. Magee, eds.), John Wiley & Sons, New York, pp. 307-388.

68. M. Dole, *The Radiation Chemistry of Macromolecules* (M. Dole, ed.), Academic Press, New York, N.Y. 1972, Vol. 1, p. 340.

69. T. R. Waite, *Phys. Rev.*, *107*, 463 (1957).

70. A. Charlesby and S. H. Pinner, *Proc. Roy. Soc.*, (London), *A269*, 367 (1959).

71. H. Wesslau, *Makromol. Chem.*, *20*, 111 (1956).

72. O. Saito, H. Kang, and M. Dole, *J. Chem. Phys.*, *46*, 3607 (1967).

73. H. Y. Kang, O. Saito, and M. Dole, *J. Polymer Sci.*, Part C, No. *25*, 123 (1968).

74. F. D. Swanson and J. L. Madsen, paper presented at National Sample Technical Conference, Huntsville, Alabama, Oct. 1971.

Chapter 9

DIGITAL SPECTRAL PEAK RESOLUTION

Lee E. Vescelius

The Firestone Tire and Rubber Company
Central Research Laboratories
Akron, Ohio

I. INTRODUCTION

An absorption spectrum consisting of overlapping peaks, each
with different intensity, position and shape, contains more information than can be obtained by visual inspection. This article
presents a digital computer program (CURVE) for resolution of these
spectral component peaks. Gaussian and Lorentzian peaks can be
added in any combination with free variation of the position,
height, and width of each component. Similar programs [1,2] have
been described in the literature earlier. It is hoped that this
article will prove useful to workers in various fields because it

provides easily accessible FORTRAN programs, comments on their use,
and compares them to an alternate analog method of curve resolution,
and because the program is very flexible and adaptable to different
problems.

Unresolved spectral peaks occur in every area of spectroscopy.
The original motivation for this program was the need to determine
the amount of block styrene in a styrene-butadiene copolymer [3].
When styrene monomer units occur in a series uninterrupted by butadiene units, the environment of the styrene or the proton is altered
significantly. As a result, the nuclear magnetic resonance peak of
this proton is shifted to lower frequencies than that of the same
proton on an isolated styrene unit. The relative area of this over-
lapped peak leads directly to the desired information about the
sequence of styrene and butadiene units.

Similar problems arise in the many areas of chromatography.
For example, the gel permeation chromatogram of a partially coupled
polymer sample will show a bimodal or more complicated pattern.
Resolution of the chromatogram components can give information on
the efficiency and mechanism of the coupling reaction.

II. METHOD

Program CURVE is a simple numerical least-squares curve-fitting
digital computer program. The function to be minimized is S^2, the
sum of the squares of the differences of the experimental composite
curve $Y(x_i)$ and the mathematical composite curve $y(x_i)$ both
evaluated at the points x_i:

$$S^2 = \sum_{i=1}^{M} [Y(x_i) - y(x_i)]^2 \tag{1}$$

where M is the number of input data coordinates. The curve $y(x_i)$
is the sum of any desired number (NG) of Gaussian functions $G_j(x_i)$
and any desired number (NL) of Lorentzian functions $L_j(x_i)$

$$y\,(x_i) = \sum_{j=1}^{NG} G_j\,(x_i) + \sum_{j=1}^{NL} L_j\,(x_i) \tag{2}$$

where

$$G_j\,(x_i) = A_j \exp\left[-4(\ln 2)\frac{(x_i - C_j)^2}{B_j}\right] \tag{3}$$

and

$$L_j\,(x_i) = \frac{A_j}{1 + 4\dfrac{(x_i - C_j)^2}{B_j}} \tag{4}$$

These specific forms were chosen so that in each case A_j is the peak height, B_j is the width at half height, and C_j is the x co-ordinate of the peak position of the component curve.

Clearly S^2 is a function of 3N variables; if N = NG + NL, then $S^2 = S^2\,(A_j,\,B_j,\,C_j)$ for j = 1 to N. The minimization of S^2 with respect to these 3N variables can be accomplished with any suitable digital minimization program. Subroutine STEPIT is used here and highly recommended because it is extremely efficient and versatile. STEPIT is held under a copyright and is not listed here but is available at a very small handling charge from the Quantum Chemistry Program Exchange as listed in Ref. 4.

It is important to be aware that there is no unique minimum solution, and many local minima will be unsatisfactory. Different starting parameters in Eq. (2) will lead to different final solu-tions. Each component curve must be placed at the beginning in such a way that it cannot wander away from its desired position. Also, the minimum number of required components should be used since the presence of one or more redundant components is likely to make all solutions unsatisfactory.

The FORTRAN listing of the computer program is included at the end of this chapter. The main program, CURVE, handles data input and output and initialization of minimization parameters. At each change of variables STEPIT calls subroutine ADD to evaluate S^2 using Eqs. (1) to (4). ADD saves as much information as possible from earlier calculations by evaluating only those component curves that have been changed by STEPIT. The tails of the component curves with amplitude less than 10^{-6} are ignored. CURVE also calls for plotting the data, but since plotter hardware and software vary so greatly from one computer ot another, the plotting subroutines are not included here.

Note in subroutine ADD that Eqs. (3) and (4), the Gaussian (GAUS) and Lorentzian (FLOR) functions, are handled in two function statements. The curve fitting program could be adapted to radically different problems by simply substituting other analytic functions. Another alternative would be to use a function subroutine to define a component of any standard shape [1], such as a skewed curve or some combination of other analytic functions, which could be scaled and positioned to minimize S^2.

The listed FORTRAN program was run on an IBM 360 Model 50 computer. Another version was run on a 16K core machine with 16-bit word length, but the standard precision was not great enough to allow a satisfactory solution.

Mochel and Claxton [5] have described a completely different solution of the spectral curve resolution problem using an analog computer. Height, width, and position of each Gaussian and Lorentzian are all controlled independently by the computer operator who views each component and the resulting spectrum on an oscilloscope. In this case S^2 is never calculated. The criterion to be satisfied is the visual fit of the resulting spectrum with a photograph of the experimental curve. The greatest advantage of the analog computer is the constant interaction of the operator with the solution. In addition, there is no need to digitize the experimental spectrum. The disadvantages are the occasional instability of the electronic components and the awkwardness of setting up the analog computer if it is not dedicated to this problem. Also, of course, digital equipment and programmers are more widely available than analog.

Nevertheless in this laboratory, where both techniques are available, the analog solution seems preferred.

III. INPUT DATA

A sample nmr spectrum is shown in Fig. 1. The data for this spectrum was digitized manually from the nmr chart. More sophisticated digitization by an on-line computer or a photodetector or other curve following technique would be more convenient. The input values for the spectrum are shown in Table 1 in the required order. Each line of data represents the image of one computer card. The first three data are the total number of spectral points entered (NP = 50), the x coordinate of the first data point (X1 = 439 Hz), and the standard increment for the x coordinates (XINC = 1). The next fifty pairs of numbers are the x,y coordinates of the experimental spectrum. The x coordinates are left blank unless the increment from the previous coordinate is different from XINC. In its present form the program will accept up to 100 coordinate pairs. The number of data points should be kept as low as possible to minimize computing time, but to get a suitable fit it is often necessary to weight the regions of greatest curvature by entering a higher density of points.

The number of Gaussian components (NG = 3) and Lorentzian components (NL = 3) are next. These can be specified in any combination up to a total of nine components. A nonzero value of the variable NTRAC can be entered here which will signal subroutine STEPIT to point out a trace map of the minimization process, but generally NTRAC is left blank.

The beginning values of the component curves are entered next, Gaussians first and then Lorentzians. Each card gives the height, width, and then peak position of one component. The units of these parameters are, of course, the same as those used to measure the x,y coordinates of the experimental spectrum.

The last 12 input data are the minimization parameters used by STEPIT. When the minimization begins, each variable is changed to

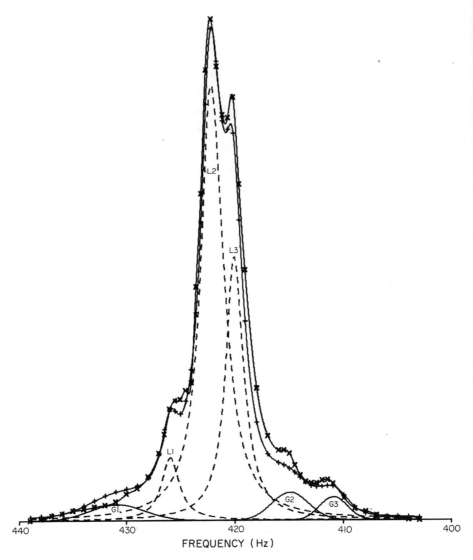

FIG. 1.   Input data (before minimization).   XX: experimental
data points connected by a smooth curve, ++: calculated spectrum,
dashed lines: Lorentzian components, solid lines: Gaussian components.

TABLE 1

Input Data

		(NP =) 50	(X1 =) 439	(XINC =) 1			
x	y	x	y	x	y	x	y
	0.0		0.6		2.2		3.1
	4.3		5.0		6.0		6.8
	8.2		11.5		14.0		19.0
	29.0	426.5	37.3	426.0	48.0	425.5	51.5
425.0	52.2	424.5	56.0	424.0	59.2	423.5	100.5
423.0	140.3	422.5	193.8	422.0	215.6	421.5	195.1
421.0	174.2	420.5	173.0	420.0	182.0	419.5	144.5
419.0	107.5		57.0		36.4		30.2
415.5	30.2	415.0	28.7	414.5	23.7	414.0	19.5
413.5	16.9	413.0	16.0	412.5	16.0	412.0	17.2
411.5	17.5	411.0	15.9		9.5		5.0
	3.8		2.2		1.2		1.0
	0.2		0.0				

(NG =) 3   (NL =) 3

	Height	Width	Position
Gaussian	7.0	6.0	431.0
	12.0	4.0	415.0
	10.0	3.0	411.0
Lorentzian	187.0	2.5	422.0
	113.0	2.0	420.0
	27.0	2.0	426.0

DAMIN			DA			AMIN			AMAX		
0.5	0.1	0.1	4.0	0.2	2.0	2.0	1.0	405.0	230.0	10.0	440.0

above and below its starting value by a given increment, DA. STEPIT
determines values for each variable increasing the increment if
necessary which straddle a minimum value of S^2 [Eq. (1)], and then
by quadratic interpolation finds the optimum value of that variable.
As STEPIT continues to cycle through all the variables, the increment for each variable is decreased because its optimum value is
known with more certainty. The minimization is considered complete
when the increment for each variable is as small as the value given
by DAMIN. At no time during the minimization will the value of a
variable be allowed to be greater than its corresponding value of
AMAX or less than AMIN.

For the parameters DAMIN (I), DA (I), AMAX (I), AMIN (I),
I = 1, 3, the subscript 1 specifies height, subscript 2 specifies
width, and subscript 3 specifies peak position for all of the component curves. The ease of altering the program to specify each
parameter for each component uniquely is obvious.

There is one more minimization parameter, MASK (I), arbitrarily
set equal to 0 in program CURVE. If used more flexibly, a nonzero
value entered for MASK (I) will instruct STEPIT to hold the corresponding variable fixed during the minimization. For instance, it
may be desirable to hold the peak position or width of any or all
of the component curves constant.

As mentioned earlier, the advantage of the analog curve resolution technique is constant interaction of the operator with the
machine. The user of the digital program CURVE can control the
result equally well, but that control is not as easy or obvious.
The tools used to exert this control are the minimization parameters
DAMIN, DA, AMAX, AMIN, and MASK, and the starting values of the component curves.

In order to determine these starting values, first estimate the
variables of each component by simply examining the experimental
spectrum, then plot these estimated components and their sum with the
experimental spectrum as shown in Fig. 1. Finally, make changes as
necessary to get a reasonable preliminary fit. It seems useful to
keep the input spectrum mostly less than the experimental spectrum.

If any one component was much larger than needed, its smaller neighbors could easily drift away from their intended positions and an undesirable local minimum would be found. When setting up the input data, keep in mind that the order in which STEPIT first tries to optimize the variables is the same as the order in which they are entered: Gaussians first and then Lorentzians. For each component curve, the order is first the height, the width, and then the peak position. By mentally changing the variables in this order, it is easy to avoid having any component drift far away from its best position.

The minimization parameters could be used much more powerfully than in the example. When attacking a new problem, it might be most useful to alter the input data to assign AMIN, AMAX, and MASK for each variable of each component curve.

IV. CONCLUSIONS

Output data for the sample problem are given in Table 2. Figure 2 shows the very good fit obtained. The solution required 2573 computations by subroutine ADD. The number of computations and computer time (8.5 min) could be lowered by judiciously raising the DAMIN values.

TABLE 2

Output Data

S^2 = 200.9

Experimental area = 1338

Calculated area   = 1387

Number of function computations = 2573

	Height	Width	Position	Relative area
Gaussian	4.43	6.65	429.3	0.023
	14.72	4.15	415.4	0.047
	11.43	2.54	411.3	0.022
Lorentzian	182.61	2.33	422.1	0.482
	135.19	2.40	420.0	0.367
	26.83	1.94	425.8	0.059

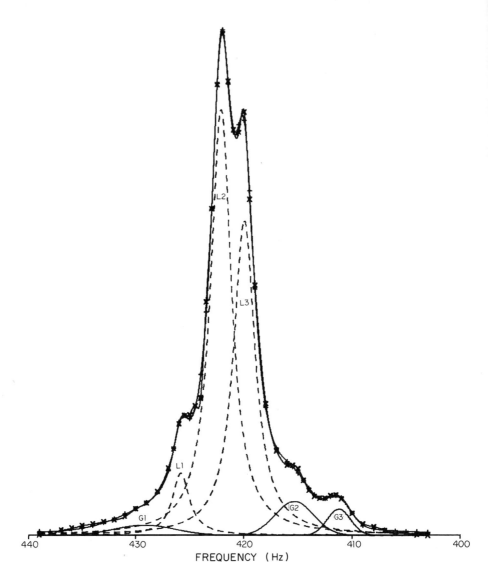

FIG. 2. Output data (after minimization). XX: experimental data
points connected by a smooth curve, ++: calculated spectrum, dashed
lines: Lorentzian components, solid lines: Gaussian components.

APPENDIX: FORTRAN COMPUTER PROGRAM

```
      PROGRAM CURVE
      DIMENSION DAMIN(3),DA(3),AMIN(3),AMAX(3)
      DIMENSION S(27),ID(40)
      COMMON NV,NTRAC,MATRX,CHISQ,MASK(27),X(3,9),XMAX(27),XMIN(27),
     1 DLTAX(27),DLMIN(27),ERR(27,3),
     2 NG,NL,NT,NGP,NP,HNMR(100),HCALC(100),X1,XINC,H(9,100),A(1200)
      JI = 2
      JO = 5
      MATRX = 0
    1 CONTINUE
      READ (JI,105) ID
  105 FORMAT (40A2)
      READ(JI,100) NP,MX1,XINC,(A(I),HNMR(I),I=1,NP)
  100 FORMAT(I5,5X,2F10.5/(8F10.5))
      IF (NP) 99,99,15
   15 READ(JI,110) NG,NL,NTRAC
  110 FORMAT(3I5)
      NGP = NG + 1
      NT = 3*NT
      NV = 3*NT
      A(1) = X1
      DO 17 I=2,NP
      IF (A(I)) 16,16,17
   16 A(I)=A(I-1)-XINC
   17 CONTINUE
      READ (JI,120) (X(J,I),J=1,3),I=1,NT)
  120 FORMAT (3F10.5)
      DO 18 I=1,NV
      MASK(I) = 0
   18 CONTINUE
```

```
      READ (JI,115) DAMIN,DA,AMIN,AMAX
  115 FORMAT(12F5.0)
      DO 19 I=1,NV,3
      DLMIN(I) = DAMIN(1)
      DLMIN(I+1) = DAMIN(2)
      DLMIN(I+2) = DAMIN(3)
      DLTAX(I) = DA(1)
      DLTAX(I+1) = DA(2)
      DLTAX(I+2) = DA(3)
      XMIN(I) = AMIN(1)
      XMIN(I+1) = AMIN(2)
      XMIN(I+2) = AMIN(3)
      XMAX(I) = AMAX(1)
      XMAX(I+1) = AMAX(2)
      XMAX(I+2) = AMAX(3)
   19 CONTINUE
      K = 1099 + NV
      DO 219 I=1100,K
      A(I) = -10.
  219 CONTINUE
      CALL STEPIT(ADD)

      WRITE(JO,170)
      WRITE(JO,180)  (HNMR(I),I=1,NP)
      WRITE(JO,180)  (HCALC(I),I=1,NP)
      DO 179 I=1,NT
      WRITE(JO,180)  (H(I,J),J=1,NP)
  179 CONTINUE
  180 FORMAT (//(10E12.4))
      STOT = 0.0
      SCALC = 0.0
      IF (NG) 23,23,20
   20 DO 21 I=1,NG
   21 S(I) = X(1,I)*X(2,I)*1.064467
```

```
      IF (NL) 25,25,23
23    DO 24 I=NGP,NT
24    S(I) = X(1,I)*X(2,I)*1.5707963
25    DO 26 I=1,NT
26    SCALC = SCALC + S(I)
      DO 27 I=1,NT
27    S(I) = S(I)/SCALC
      DO 28 I=2,NP
28    STOT = STOT + (A(I-1)-A(I))*(HNMR(I) + HNMR(I-1))
      STOT = 0.5*STOT
      WRITE(JO,150) ID,STOT,SCALC
150   FORMAT(1H1,40A2//' EXPERIMENTAL AREA = ',F12.5/' CALCULATED AREA
     1= ',F14.5)
      IF (NG) 40,40,35
35    WRITE (JO,155)
155   FORMAT (///' GAUSSIAN PEAKS '/)
      WRITE (JO,160) (J,(X(I,J),I=1,3),S(J),J=1,NG)
160   FORMAT (I4,5X,'HEIGHT = ',F10.3,5X,'WIDTH = ',F10.3,5X,'POSITION ='
     1,F10.3,10X,'AREA + ',F10.3)
40    IF (NL) 50,50,45
45    WRITE (JO,165)
165   FORMAT (///' LORENTZIAN PEAKS'/)
      WRITE (JO,160) (J,(X(I,J),I=1,3),S(J),J=NGP,NT)
50    DO 60 I=1,NP
      DO 55 K=1,NT
      J = I + K*NP
55    A(J) = H(K,I)
      J = I + (NT+1)*NP
      A(J) = HCALC(I)
      J = I + (NT+2)*NP
      A(J) = HNMR(I)
```

```
 60 CONTINUE
    WRITE (JO,170)
170 FORMAT (1H1)
    CALL LPLT(A,NP,NT+3,0)
    WRITE(JO),170)
    GO TO 1
 99 CALL EXIT
    END
    SUBROUTINE ADD
    DIMENSION I1(9),I2(9),ISAV(9)
    COMMON NV,NTRAC,MATRX,CHISQ,MASK(27),X(3,9),XMAX(27)XMIN(27),
   1 DLTAX(27),DLMIN(27),ERR(27,3),
   2 NG,NL,NT,NGP,NP,HNMR(100), HCALC(100),XH,XINC,H(9,100),A(1200)
    GAUS(A,B,C,X) = A*EXP(-2.77259*(X-C)/B)**2)
    FLOR(A,B,C,X) = A/(1.+4.*((X-C)/B)**2)
    DO 9 I=1,NT
    K = 1099 + (I-1)*3

  C    CALCULATE ONLY THOSE CURVES WHICH HAVE CHANGED SINCE THE LAST CALCULATION

    DO 2 J=1,3
    K = K + 1
    IF (A(K)-X(J,I)) 1,2,1
  1 ISAV(I) = 1
    IF (I-NG) 3,3,4
  2 CONTINUE
    ISAV(I) = 0
    GO TO 9

  C    CALCULATE ONLY THOSE PORTIONS OF ANY CURVE GREATER THAN 10**(-6)

  3 WIDTH = X(2,I)*SQRT((9.2+ALOG(X(1,I))/2.77259)
    GO TO 5
```

```
    4   WIDTH = 0.5*X(2,I)*SQRT(10000.*X(1,I)-1.)
    5   J = 0
  206   J = J + 1
        IF(X(3,I)+WIDTH-A(J)) 206,207,207
  207   I1(I) = J
        DO 208 J1 = N,NP
        IF(X(3,I)-WIDTH-A(J1)) 208,209,209
  208   CONTINUE
        J1 = NP + 1
  209   I2(I) = J1 - 1

    9   CONTINUE

   10   IF (NG) 30,30,10
        DO 20 J=1,NG
        IF (ISAV(J)) 20,20,12
   12   X1 = X(1,J)
        X2 = X(2,J)
        X3 = X(3,J)
        J1 = I1(J)
        J2 = I2(J)
        DO 12 I=1,J1
   13   H(J,I) = 0.0
        DO 14 I=J2,NP
   14   H(J,I) = 0.0
        DO 15 I=J1,J2
        H(J,I) = GAUS(X1,X2,X3,A(I))
   15   CONTINUE
   20   CONTINUE
   30   IF (NL) 60,60,30
        DO 50 J=NGP,NT
```

```
      IF (ISAV(J)) 50,50,35
   35 X1 = X(1,J)
      X2 = X(2,J)
      X3 = X(3,J)
      J1 = I1(J)
      J2 = I2(J)
      DO 38 I=1,J1
   38 H(J,I) = 0.0
      DO 39 I=J2,NP
   39 H(J,I) = 0.0
      DO 40 I=J1,J2
   40 H(J,I) = FLOR(X1,X2,X3,A(I))
   50 CONTINUE
   60 CHISQ = 0.0
      DO 80 I=1,NP
      HCALC(I) = 0.0
      DO 70 J=1,NT
   70 HCALC(I) = HCALC(I) + H(J,I)
   80 CHISQ = CHISQ + (HNMR(I)-HCALC(I))**2
      K = 1099
      DO 90 I=1,NT
      DO 90 J=1,3
      K = K + 1
   90 A(K) = X(J,I)
   95 CONTINUE
      RETURN
      END
```

REFERENCES

1. W. D. Keller, T. R. Lusebrink, and C. H. Sederholm, *J. Chem. Phys.*, *44* (2), 782-793 (1966).

2. H. Stone, *J. Optical Soc. Amer.*, *52* (9), 998-1003 (1962).

3. V. D. Mochel, *Macromolecules, 2,* 537 (1969).

4. Subroutine STEPIT, (copyright by J. P. Chandler), Program Number 66.1, Quantum Chemistry Program Exchange, Indiana University, Bloomington, Indiana.

5. V. D. Mochel and W. E. Claxton, *J. Polymer Sci.*, A-1, *9,* 345-362 (1971).

Chapter 10

COMPUTER CALCULATIONS OF LIGHT SCATTERING FROM CRYSTALLINE POLYMERS

Ashok Misra

Research Department
Monsanto Polymers and Petrochemicals Co.
Springfield, Massachusetts

Richard S. Stein

Polymer Research Institute
University of Massachusetts
Amherst, Massachusetts

I. INTRODUCTION

Light-scattering [1-3] calculations from crystalline polymers
are carried out using two approaches: (1) the model approach and
(2) the correlation function approach. The model approach describes
the system in terms of discrete scattering units such as anisotropic
spheres, disks, or rods over which the scattering amplitude is summed. The calculations are then modified to account for deviations
from ideality. In the correlation function approach, the scattering
is described in terms of correlation functions relating the separation of pairs of volume elements to the correlation in their average
refractive index, anisotropy, and orientation of their optic axes.
Since crystalline polymers generally have a spherulitic superstruc-
ture, greater stress is given to scattering from spherulites which
are approximated by spheres or disks. Scattering from other shapes
is also discussed.

Most of the theories developed to account for the observed results cannot be treated analytically and are, consequently, evaluated by computer simulation techniques. For example, spherulites
are generated by the computer in which the position of spherulite
centers and the orientation of optic axes are assigned statistically and the scattering amplitudes are evaluated by numerically summing contributions from all volume elements. Such calculations have
to be repeated several times in order to obtain an average scattering pattern.

II. SCATTERING FROM SPHERULITES

A. The Model Approach

In the model approach the scattering region of the polymer is
represented by an idealized structure and the scattering pattern is
calculated from this structure. An example of such a calculation
is the representation of the scattering from a spherulitic polymer
by that from an isolated anisotropic sphere [4,5] or a disk [5-8].
The scattering from such a uniform sphere may be calculated analytically. The theoretical predictions differ from experimental

results because of complications arising from several factors such
as interspherulitic interference, truncation among spherulites, internal disorder within the spherulites and multiple scattering.
The calculations are thus modified to account for deviations from
ideality in actual systems. Light-scattering patterns may also be
calculated from other structures such as anisotropic sheafs or rods.
The availability and the speed of computation by computers in the
last decade has contributed greatly to the progress made in calculating light-scattering patterns and in the development of modified
light-scattering theories.

The theory of light scattering by spherulites was first developed by considering the spherulite as a homogeneous anisotropic
sphere [4] in which the optic axis is located along or perpendicular to the radius of the sphere. The scattering is described in
terms of the coordinate system shown in Fig. 1. The vertical direction is taken as the z-axis and the incident beam propagates in the
x-direction. The scattering element has radial and tangential
polarizabilities, α_r and α_t, respectively, and is located in the
spherulite by the polar coordinates r, α, and Ω. α_s is defined as
the polarizability of the surrounding medium. The direction of the
scattered ray is described in terms of the conventional scattering
angle θ and the azimuthal angle μ. The scattered intensity I is
found by averaging the square of the scattering amplitude E which
is found by summing the amplitude contributions from all of the
volume elements of the scattering objects (the anisotropic spherulite). E can be expressed as

$$E = \sum_n E_n \qquad (1)$$

where E_n is the scattering amplitude contributed by the nth volume
element.

In practice, the sum of Eq. (1) is generally approximated by
an integral over the bounds of the scattering particle and is given
by the expression

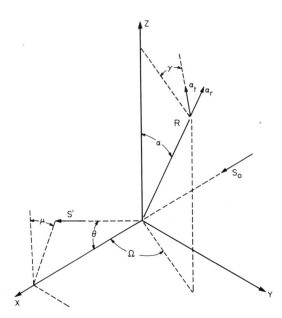

FIG. 1.   The coordinate system used in describing the scattering from an anisotropic sphere [from R. S. Stein and M. B. Rhodes,
J. Appl. Phys., 31, 1873 (1960), Fig. 5].

$$E = K_1 \int (\vec{M} \cdot \vec{o}) \, \exp[ik(\vec{r} \cdot \vec{s})] \, d\vec{r} \qquad (2)$$

where $\vec{M}$ is the induced dipole moment in the scattering element at
a distance r from the center of the spherulite; $\vec{o}$ is a unit vector
perpendicular to the scattered ray and in the plane of polarization
of the light transmitted by the analyzer; $k = 2\pi/\lambda$ is the wave number in the medium (is the wavelength); and $\vec{s}$ is the propagation
vector $(\vec{s}_o - \vec{s}')$, where $\vec{s}_o$ and $\vec{s}'$ are unit incident and scattered
beam vectors.   In terms of the coordinator system in Fig. 1,

$$\vec{r} = r \, [(\sin \alpha \cos \Omega)\vec{i} + (\sin \alpha \sin \Omega)\vec{j} + (\cos \alpha)\vec{k}] \qquad (3)$$

$$\vec{s} = (1 - \cos \theta)\vec{i} - (\sin \theta \sin \mu)\vec{j} - (\sin \theta \cos \mu)\vec{k} \qquad (4)$$

$$d\vec{r} = r^2 \sin \alpha \, d\alpha \, d\Omega \, dr$$

The induced dipole moment $\vec{M}$ is given as

$$\vec{M} = (\alpha_r - \alpha_t)(\vec{E}_o \cdot \vec{a}_r)\vec{a}_r + \alpha_t \vec{E} \tag{6}$$

where $\vec{E}_o$ is the incident field vector and $\vec{a}$ is a unit vector along the radial direction.

The vector O for H_V (polarizer and analyzer perpendicular) and V_V (polarizer and analyzer parallel) are respectively given as [8]

$$\vec{O}_{H_V} = -\sin \rho_2 \, \vec{i} + \cos \rho_2 \, \vec{j} \tag{7}$$

$$\vec{O}_{V_V} = -\sin \rho_1 \, \vec{i} + \cos \rho_1 \, \vec{k} \tag{8}$$

where ρ_1 and ρ_2 are defined as

$$\cos \rho_1 = \frac{\cos \theta}{(\cos^2\theta + \sin^2\theta \, \cos^2\mu)^{1/2}} \tag{9}$$

$$\cos \rho_2 = \frac{\cos \theta}{(\cos^2\theta + \sin^2\theta \, \sin^2\mu)^{1/2}} \tag{10}$$

Scattering amplitudes can now be determined by substituting Eqs. (3) through (6) and (7) or (8) for H_V or V_V polarization in Eq. (2) and integrating. The respective intensities are then dound by squaring the amplitude and are given as [4,5,8]

$$I_{H_V} = K_2 \left\{ V_s \cos \rho_2 \left(\frac{3}{U^3}\right) \left[(\alpha_r - \alpha_t)\frac{\cos^2(\theta/2)}{\cos \theta} \sin \mu \, \cos \mu \right. \right.$$

$$\left. \left. X \, (4 \sin U - U \cos U - 3 \, Si \, U) \right] \right\}^2 \tag{11}$$

$$I_{V_V} = K_2 \left\{ V_s \cos \rho_1 \left(\frac{3}{U^3}\right) \left[(\alpha_t - \alpha_s)(2\sin U - U \cos U - Si \, U) \right. \right.$$

$$+ (\alpha_r - \alpha_s)(Si \, U - \sin U) - (\alpha_t - \alpha_r) \frac{\cos^2(\theta/2)}{\cos \theta} \cos^2 \mu$$

$$\left. \left. X \, (4 \sin U - U \cos U - 3 \, Si \, U) \right] \right\}^2 \tag{12}$$

where K_2 is a physical constant, and $V_s = 4\pi R_s^3 / 3$ is the volume of

the spherulite with R_s being the spherulite radius. U is the shape
factor and for a sphere it is;

$$U = \frac{4\pi R}{\lambda} \sin \frac{\theta}{2} \tag{13}$$

Si U is the sine integral defined as

$$Si\ U = \int_0^U \frac{\sin x}{x}\ dx \tag{14}$$

which is solved as a series expansion sum for computation purposes.

Complete light-scattering patterns can now be easily obtained
by calculating I_{H_v} and I_{V_v} for several values of the scattering
angle θ and the azimuthal angle μ. Samuels [5] has utilized the
Calcomp plotter attachment to obtain complete patterns (Fig. 2)
directly from the computer. The H_v intensity depends upon the anisotropy $(\alpha_r - \alpha_t)$ of the spherulite and varies with $\sin^2\mu \cos^2\mu$
which gives a four-leaf clover pattern with scattering maxima
occurring at odd multiples of $\mu = 45°$. The scattering angle at
these maxima is a function of the spherulite radius given as

$$U_{max} = 4\pi \frac{R}{\lambda} \sin \frac{\theta_m}{2} = 4.1 \tag{15}$$

Higher-order scattering maxima in θ are also predicted but are
usually washed out by an averaging effect if scattering is calculated for a system with a distribution of spherulite sizes.

The V_v pattern, on the other hand, depends upon the polarizability of the surroundings α_s as well as on the anisotropy of the
spherulite. The amplitude consists of two components, one of which
depends on α_s but is angularly independent, while the other depends
on $(\alpha_r - \alpha_t)$ and $\cos^2\mu$ giving rise to twofold symmetry.

Experimentally it has been observed that the overall V_v intensity goes through a maximum during crystallization [4] while the
corresponding H_v intensity increases monotonically. A qualitative
explanation of such an observation is quite straightforward. The
H_v scattering at small angles depends only upon fluctuations in the

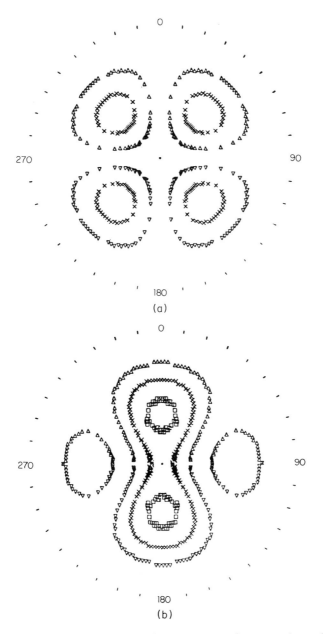

FIG. 2. Theoretical scattering patterns from a spherulite:
(a) H_V, (b) V_V [from R. J. Samuels, *J. Polymer Sci.* A-2, *9*, 2165
(1971), Fig. 7].

magnitude and optic axis orientation of anisotropic regions, whereas the V_V scattering also depends upon fluctuations in the density
or average polarizability. The V_V intensity maximum arises from
the latter contribution. The origin of this average polarizability
fluctuation is the difference between the average polarizability of
the spherulite and that of its surrounding material. Its contribution is greatest when the polymer is about half spherulitic after
which it decreases due to interference as the spherulites become
volume filling. Thus the V_V scattering maximum should occur when
the fraction of spherulites, ϕ_s approaches 1.0. The residual
scattering at this time arises from anisotropic contribution. A
second increase in V_V intensity and a continued increase in H_V intensity after ϕ_s reaches unity is associated with a further increase
in anisotropy as a consequence of secondary crystallization.

A simpler two-dimensional analog of the above theory has also
been developed [7,8]. It consists of a homogeneous, anisotropic,
infinite thin disk, placed in a plane perpendicular to the incident
beam shown in Fig. 3. Expressions for calculating scattering intensities from disks may be developed following the procedure

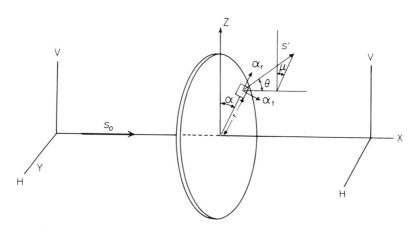

FIG. 3. Coordinate system for scattering from an anisotropic
disk [from C. Picot, R. S. Stein, M. Motegi, and H. Kawai, *J.
Polymer Sci.* A-2, *8*, 2115 (1970), Fig. 2].

outlined earlier and are given as

$$I_{H_V} = K_3 \left\{ \cos \rho_2 A \frac{1}{w^2} (\alpha_r - \alpha_t) [2(1 - J_0(w)] \right.$$
$$\left. \times \sin \mu \cos \mu \right\}^2 \qquad (16)$$

$$I_{V_V} = K_3 \left\{ \cos \rho_1 A \frac{2}{w^2} [(\alpha_t - \alpha_s)wJ_1(w) + (\alpha_r - \alpha_t) [1 - J_0(w)] \right.$$
$$\left. - (\alpha_r - \alpha_t)\cos^2\mu\{2[1 - J_0(w)] - wJ_1(w)\}] \right\}^2 \qquad (17)$$

where K_3 is a physical constant, w is a reduced variable $2\pi R \sin$
(θ/λ), A is the disk area, and $J_0(w)$ and $J_1(w)$ are respectively the
zero-order and first-order Bessel functions. It has been found
that as long as the plane of the disk makes an angle of 90° with
the incident beam, the shape and relative intensity of the scattering pattern is very similar to that obtained with the sphere calculation. Thus, the modifications of the theory are generally
studied using the simpler two-dimensional approach and the results
can be applied to the interpretation of the physically more realistic three-dimensional case for which the computational times would
be prohibitively long.

The effect of the optic axis tilting on the spherulite-scatter-
ing pattern can be analyzed for the two-dimensional case simply by
replacing μ by $(\mu + \beta)$, β being the angle between the optic axis
and the spherulite radius. Normal types of H_V patterns with the
lobes at $\mu = 45°$ are obtained when the optic axis lies at an angle
β close to 0° or 90° to the radius; while the lobes occur along the
polar directions when the axis is tilted with β close to 45°. Examples of experimental H_V and V_V patterns along with corresponding
photomicrographs for cases where $\beta = 0°$ or 90° and where $\beta = 45°$
are presented in Figs. 4 and 5, respectively.

Theoretical scattering patterns from isolated spherulites are
in qualitative agreement with the shape of experimentally obtained
patterns except that the latter arc of a "speckled" nature which
can be explained on the basis of interspherulitic interference. A
quantitative comparison further shows the theory to differ from

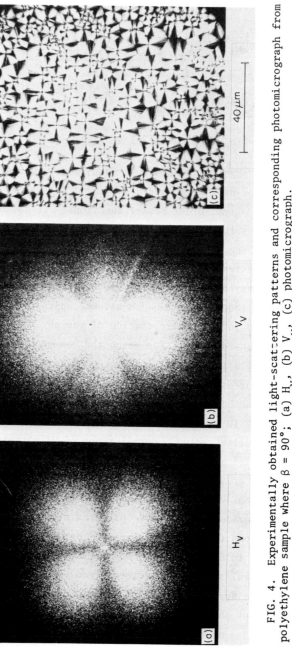

FIG. 4. Experimentally obtained light-scattering patterns and corresponding photomicrograph from a polyethylene sample where β = 90°; (a) H_V, (b) V_V, (c) photomicrograph.

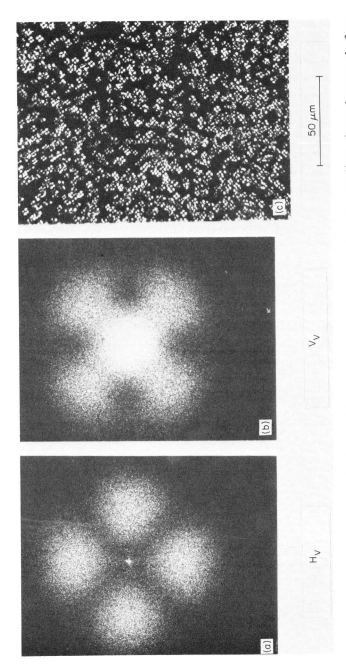

FIG. 5. Experimentally obtained light-scattering patterns and corresponding photomicrograph from a polybutylene terephthalate sample where $\beta = 45°$; (a) H_V, (b) V_V, (c) photomicrograph.

experiment in several respects: (1) the theory predicts zero scattering intensity at $\theta = 0°$, whereas a finite intensity if found experimentally; (2) the theory predicts zero scattering intensity along
azimuthal angles of $\mu = 0°$ and $90°$, whereas a finite intensity is observed experimentally; (3) the theoretical patterns show a much more
rapid decrease in intensity at large scattering angles than do experimental patterns. These differences can generally be attributed
to several factors such as truncations arising from impingment of
spherulites, disorder in the internal structure of spherulites and
multiple scattering. Modifications are now required in the relatively simple theory of scattering to account for the deviations from the
ideal isolated homogeneous spherulite.

B. Interspherulitic Interference

The results from the theory for the scattering from isolated
spherulites are in general agreement with the experimental patterns
accounting qualitatively for most of the observed features. It
should be noted that this is an over-simplification since only isolated spherulites are considered in the theory, whereas the actual
systems can deviate because of interspherulitic interference and
truncation arising from neighboring spherulites. The theory of interference among small groups of spherulites was first worked out by
Stein and Picot [9] and latter extended to a larger number of spherulites by Stein and Prud'homme [10]. Considering a system containing N spherulites, the vector $\vec{r}$ of Eq. (2) can be defined as (Fig. 6)
[9].

$$\vec{r} = \vec{R}_i + \vec{r}_i \tag{18}$$

The center of spherulite I is arbitrarily taken as the origin, thus
$\vec{R}_i = 0$ for the first particle. Substituting Eq. (18) in (2) one gets

$$E = K_4 \left\{ \sum_{i=1}^{n} \int_{r_i} [\vec{M}(r_i) \cdot \vec{o}] \exp[\vec{R}_i + \vec{r}_i) \cdot \vec{s}] \, d\vec{r}_i \right.$$
$$\left. + \int_{r_s} (\vec{M}_s \cdot \vec{o}) \exp[ik(\vec{r}_s \cdot \vec{s}) \, d\vec{r}_s] \right\} \tag{19}$$

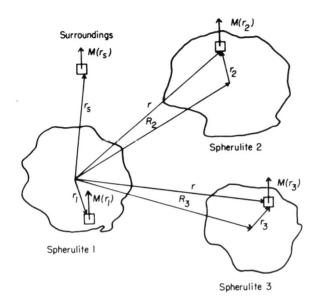

FIG. 6. Different regions contributing to the scattering from a system of several particles in isotropic homogeneous surroundings [from R. E. Prud'homme and R. S. Stein, *J. Polymer Sci.* A-2, *1357*, (1973), Fig. 1].

If all the particles are identical, that is, $E_1 = E_2 = E_N$, the total scattering intensity can be expressed as [10]

$$I = K_5 \left\{ E_1^2 \left[N + \sum_{i \neq j} \sum \cos k \, (\vec{r}_{ij} \cdot \vec{s}) \right] \right\} \tag{20}$$

The first term in Eq. (20) represents the scattering from isolated particles, while the second term represents the contribution due to interspherulitic interference. Results show that the general shape of the scattering pattern remains the same, but the pattern becomes "speckled" in nature as is experimentally observed from spherulitic polymers. Picot et al. [11] have experimentally demonstrated this interference effect by using groups of spherulitic starch granules or anisotropic polystyrene spherulites. As the number of spherulites in the incident beam increases, the interference effect becomes smaller and averages out to zero in the limiting case. This can be shown by writing Eq. (20) in the form:

$$I = K_5 \left\{ E_1^2 \left[N + \sum_{i \neq j} \overline{\cos k \, (\vec{r}_{ij} \cdot \vec{s})} \right] \right\}$$ (21)

where the bar denotes the average value. Assuming that the spherulitic centers are located at random, the cosine term will average
out to zero. Thus, in practice, as the number of spherulites in
the incident beam increases, the pattern becomes less speckled [10].

C. Truncation Effect

In spherulitic crystalline polymers the spherulites are generally of a volume filling nature, i.e., they fill up the medium
completely. Due to the close packing the spherulites are no longer
circular but appear like polygons (Fig. 7). Stein and Picot [12]
have modified the theory to account for the deviation in scattering

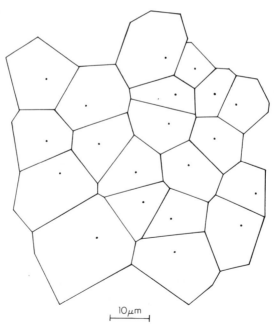

FIG. 7. A schematic representation of an array of truncated
two-dimensional spherulites.

due to truncation. This is basically done by considering the truncated spherulite to be a part of a complete disk and then subtracting the amplitude E' due to the excess area from the amplitude E_{CD}
due to the complete disk. The amplitude E_{TD} corresponding to the
truncated disk can now be written as

$$E_{TD} = E_{CD} - E'$$ (22)

Using simple truncations Stein and Picot have shown that truncation
results in a slight decrease of intensity at θ_{max}, while an increase in intensity at angles lower as well as higher than the
scattered maximum. Prud'homme and Stein [13] have considered more
complex truncations and have described the degree of truncation by
an arbitrarily defined "truncation parameter." They have shown
that the deviations from ideality increase with increasing values
of the truncation parameter. Kawai et al. [14] have developed a
more general theory of deviations from circular shapes of two-
dimensional spherulites giving equations of the form

$$I_{H_v} = \frac{K_6 \delta^2 R_o^4}{4} [A_1(\theta) - A_2(\theta)\cos 4\mu]$$ (23)

where R_o is some characteristic particle radius, and δ is the anisotropy of the spherulite $(\alpha_t - \alpha_r)$; $A_1(\theta)$ and $A_2(\theta)$ are coefficients which depend on the external shape of the disk. This theory
is more complicated but can be applied to any shape and is not restricted to the special case of truncated spherulites.

A special case is that of an incomplete spherulite generally
referred to as a sheaf. The idealized sheaf has been represented
by a fan model as shown in Fig. 8. Picot et al. [15] have calculated scattering intensities from sheaves for several apex angles
of the fan. They show that the incompleteness of the spherulites
leads to higher intensities at angles smaller as well as greater
than maximum. The effect increases with decreasing apex angle; in
the limiting case when this angle is very small, the scattering
maxima is barely observed. Misra and Stein [16] have shown that

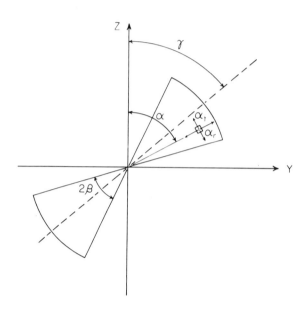

FIG. 8. The fan model of a two-dimensional sheaflike incomplete spherulite [from C. Picot, R. S. Stein, M. Motegi, and H. Kawai, *J. Polymer Sci*, A-2, *8,* 2115 (1970), Fig. 2].

the patterns observed during the early stages of the crystallization of polyethylene terephthalate are similar to those predicted by the fan (or sheaf) model. As crystallization proceeds the spherulites become more complete and the patterns change to those corresponding to spherulitic scattering.

D. Disorder Effect

First of all it should be pointed out that, in the absence of any disorder, the internal structure of a spherulite has little effect on scattering. This was shown [17] by considering an assembly of anisotropic rods imbedded in an isotropic matrix and arranged in a spherulitic distribution such that the optic axis of the rods was along or perpendicular to the radius. It was found that as long as the dimension of the rods was small compared to the wavelength of light, the scattering from such an assembly was identical with that of an isolated spherulite. However, disorder in the internal

structure of a spherulite affects the shape of the scattering pattern and helps explain the excess scattering observed experimentally.

In order to explain the effect of disorder, models can be developed by considering a heterogeneous spherulite with fluctuations
in the internal structure. Stein et al. [18] calculated scattering
intensities by adding contributions from a perfect spherulite and
those arising from density fluctuations due to internal heterogeneity. This accounted for some of the excess intensity at high
scattering angles but was not in complete agreement with the experimental observations. Later Keijzers et al. [19] considered a
spherulite which contained a certain number of perfectly spherulitic
crystallites and a certain number of randomly oriented crystallites.
The scattering was then calculated by adding weighted contributions
from the two structures in order to fit the experimental data. This
theory successfully described scattering from polymer samples which
have low spherulitic order but not from those which are highly
ordered. The shortcoming probably lies in the assumption that the
disorder orientation fluctuation is random and that no interference
between scattering from the ordered and disordered contributions is
considered.

A more specific model for an anisotropic spherulite with fluc-
uations in the optic axis orientation was presented by Stein and
Chu [20]. The orientation of the principal polarizability axis
(denoted by the unit vector $\vec{a}$) of a uniaxial crystal with respect
to the spherulite radius is described by angles β and ω defined in
Fig. 9. The scattered amplitude for H_V scattering for a two-dimensional isolated spherulite is given by [8,20]

$$E_{H_V} = \left(\frac{1}{2}\right) K_7 \; (\alpha_r - \alpha_t) \int_{r=0}^{R} \int_{\alpha=0}^{2\pi} \left\{ \cos \rho_2 \; [(\cos^2\beta - \sin^2\beta \cos^2\omega) \sin 2\alpha \right.$$

$$+ \sin 2\beta \cos \omega \cos 2\alpha] + \sin \rho_2 (\sin^2\beta \sin 2\omega \sin \alpha$$

$$\left. - \sin 2\beta \sin \omega \cos \alpha) \right\} \cos [k(\vec{r} \cdot \vec{s})] \; d\alpha \; r \; dr \qquad (24)$$

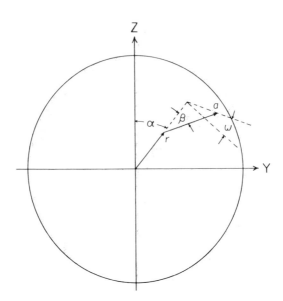

FIG. 9. Coordinates of the principal polarizability axis for
a two-dimensional spherulite [from R. S. Stein and W. Chu, *J. Polymer Sci.*, A-2, *8*, 1137 (1970), Fig. 1].

All the terms in this equation have been previously defined. For
the special case where the optic axis is confined to the plane of
spherulite, ω is zero, and Eq. (24) reduced to

$$E_{H_v} = \left(\frac{1}{2}\right) K_7 (\alpha_r - \alpha_t) \cos \rho_2 \int_{r=0}^{R} \int_{\alpha=0}^{2\pi} (\cos 2\beta \sin 2\alpha$$

$$+ \sin 2\beta \cos 2\alpha) \cos[k\ (r \cdot s)]\ d\alpha\ r\ dr \qquad (25)$$

The disorder is described by the fluctuation of the optic axis
angle β in the radial as well as the angular direction and at any
given location within the spherulite it may be expressed as

$$\beta\ (r,\ \alpha) = \beta_0 + \Delta(r,\ \alpha) \qquad (26)$$

The value of the fluctuation $\Delta(r,\ \alpha)$ at a particular volume element
depends upon the value at neighboring volume elements and can be
described in terms of Markovian statistics. Equation (25) can now

be integrated by considering the variation of β. These multiple
integrals cannot be evaluated analytically and must be evaluated by
numerical integration on a computer. However, prohibitively long
computation times are required if variations in β with r and α are
considered simultaneously. Consequently, these variations are considered separately. In the case of radial disorder, the optic axis
orientation fluctuates only with respect to the radius and Eq. (25)
reduces to

$$E_{H_V} = \pi K_7 (\alpha_r - \alpha_t) \cos \rho_2 \int_{r=0}^{R} \sin[2\mu + 2\beta(r)] \ J_2(x) \ r \ dr \qquad (27)$$

where $x = kr \sin \theta$. In the case of angular disorder, the optic
axis orientation fluctuates with the angle α and Eq. (25) reduces
to

$$E_{H_V} = \frac{1}{2} K_7 (\alpha_r - \alpha_t) \cos^2 \rho_2 \int_{\alpha=0}^{2\pi} \sin[2\beta(\alpha) + 2\alpha] \ F(\gamma) \ d\alpha \qquad (28)$$

where $\gamma = \alpha - \mu$.

$$F(\gamma) = \frac{\cos(w \cos \gamma) - 1}{w \cos \gamma} + \frac{\sin(w \cos \gamma)}{w \cos \gamma} \qquad (w = kR \sin \theta)$$

Hashimoto and Stein [21] extended the above model to consider the
effect of fluctuation in the anisotropy ($\delta = \alpha_r - \alpha_t$) expressed as

$$\delta(r, \alpha) = \delta_0 + \Delta(r, \alpha) \qquad (29)$$

As before the calculations are divided by considering radial and
angular disorder separately.

   Calculations show that in both the cases the radial disorder
contributes primarily to excess scattering at angles larger than
the maximum, while angular disorder contributes primarily to excess
scattering at angles lower than the maximum. Experimentally en-
hanced intensities are observed in both regions thus indicating a
combination radial and angular disorder.

   Recently Stein and Yoon [22] have developed a lattice theory

for light scattering from disordered spherulites. The two-dimensional spherulite is represented as a circular lattice of cells of
equal area as shown in Fig. 10. The optic axis orientation in the
i^{th} cell is given by β_i which deviates from the average value β_0
by $\Delta\beta$.

$$\beta = \beta_0 + \Delta\beta_i \tag{30}$$

For the case of $\omega = 0°$ and $\beta_0 = 90°$, the scattering amplitudes for
H_V and V_V polarizations respectively can be written as

$$E_{H_V} = \left(\frac{1}{2}\right) K_7 (\alpha_r - \alpha_t) \cos \rho_2 \sum_i \sin[2(\alpha_i + \Delta\beta_i)]$$

$$X \exp[i\, k(\vec{r}_i \cdot \vec{s})] \tag{31}$$

and

$$E_{V_V} = K_7 \cos \rho_i \sum_i [(\alpha_r - \alpha_t) \sin^2(\alpha_i + \Delta\beta_i) + \alpha_t - \alpha_s]$$

$$X \exp[i\, k(\vec{r}_i \cdot \vec{s})] \tag{32}$$

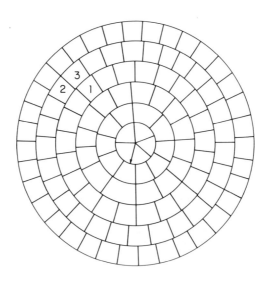

FIG. 10.   Circular lattice representation of a two-dimensional
spherulite [from D. Y. Yoon and R. S. Stein, *J. Polymer Sci.*, A-2,
12, 763 (1974), Fig. 3].

where $\vec{r}_i \cdot \vec{s} = r_i \sin\theta \cos(\mu - \alpha_i)$. The scattering intensity is thus given by

$$I = C \, (\vec{E} \cdot \vec{E}*) \tag{33}$$

where $\vec{E}*$ is the complex conjugate of $\vec{E}$.

The circular lattice is filled progressively starting from the innermost cell which is assigned an arbitrary value for its orientation. The cells in each circle are filled in a clockwise (or counterclockwise) manner until all cells are filled. The assignment of the orientation angle deviation $\Delta\beta_i$ is done in consideration to its correlation with the two nearest neighbors. The angular differences $\Delta\beta_i$ are quantized in units of an arbitrarily defined increment $\pm \, \delta$. Once the orientation in each of the lattice cells has been assigned, the scattering intensities can be calculated from Eqs. (31), (32), and (33). The disorder parameter δ can be varied in order to fit the experimental data. The theory quite successfully accounts for the deviations in the scattering from perfect spherulites.

E. Multiple Scattering

The theories for light scattering developed so far assumed that there was only one layer of spherulites in the path of the incident beam. This is not true for scattering from thin polymer films which generally contain several layers of spherulites giving rise to secondary [23] or multiple scattering [24,25]. As an example, a 25 μm-thick film will contain about five layers of spherulites which have an average diameter of 5 μm. Multiple scattering tends to make the scattering envelope more diffuse, reducing the intensity in the high-intensity regions and increasing it in the low-intensity regions. In the limiting case of very thick films, the pattern becomes circularly symmetric showing no intensity maxima. Theory for multiple scattering is based on a model made up of several layers of scattering entities as shown in Fig. 11 and helps correct measured intensities.

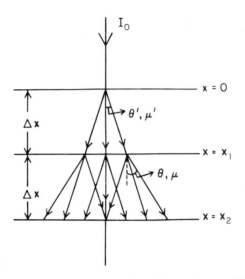

FIG. 11. Layered structure of a thin polymer film [from R. E.
Prud'homme, L. Bourland, R. T. Natarajan, and R. S. Stein, J. Polymer Sci., A-2, accepted for publication, Fig. 3].

III.   CORRELATION FUNCTION APPROACH

Debye and Bueche [26] were the first to propose a statistical
theory for scattering from a medium with density fluctuations alone
and showed that the supermolecular order in an amorphous polymer
could be described by a density correlation function $\gamma(r)$.   They
started with the equation for the amplitude of scattering

$$E = K_8 \int \rho(\vec{r}) \cdot \exp[k(\vec{r} \cdot \vec{s})] \, d\vec{r} \tag{34}$$

which is similar to Eq. (2), where the term $\rho(\vec{r})$ is the scattering
power of the medium and fluctuates from its average value ρ_0 by an
amount $\rho(\vec{r})$ so that

$$\rho(\vec{r}) = \rho_0 + \eta(\vec{r}) \tag{35}$$

For spherical symmetry the scattering intensity can be written as

$$I = K_9 \langle \eta^2 \rangle \int_{r=0}^{\infty} \gamma(r) \frac{\sin hr}{hr} r^2 \, dr \tag{36}$$

where $\langle \eta^2 \rangle$ is the mean-square fluctuation, $h = 4\pi/\lambda \sin(\theta/2)$, (r) the correlation function is defined as

$$\gamma(r) = \frac{\langle \eta_i \cdot \eta_j \rangle_r}{\langle \eta^2 \rangle} \tag{37}$$

where η_i and η_j are the fluctuations in two volume elements and $\langle \eta_i \eta_j \rangle_r$ represents an average overall pairs of volume elements separated by a fixed scalar distance r. The angular dependence of scattering is determined by the nature of $\gamma(r)$. Conversely, this function may be determined from a Fourier inversion of the variation of measured intensity with scattering angle.

The Debye-Bueche theory was extended by Goldstein and Michalek [27] to anisotropic systems involving correlations in both density and anisotropy. This theory is very general, and due to its complexity it is not readily applicable. A less general but more easily applied theory of random orientation correlations in aniso-tropic systems was proposed by Stein and Wilson [28]. They relate the scattering power to the induced dipole moment M as

$$\rho(\vec{r}) = C \ (\vec{M} \cdot \vec{O}) \tag{38}$$

Assuming that the anisotropic volume elements have cylindrical sym-metry with principal polarizabilities α_1 and α_2, the induced dipole moment $\vec{M}$ is given by

$$\vec{M} = (\alpha_1 - \alpha_2) \ (\vec{a} \cdot \vec{E}_0) \ \vec{a} + \alpha_2 \ \vec{E}_0 \tag{39}$$

where $\vec{E}_0$ is the incident electrical field vector, and $\vec{a}$ is a unit vector lying along the optic axis. The random orientation correla-tion assumes that the probability of correlation in orientation of optic axes of two volume elements depends only upon their separation and not upon the angle the optic axes make with the vector inter-connecting them. The scattering intensities can now be expressed by the equations

$$I_{H_V} = \frac{K_{10}}{15} \langle \delta^2 \rangle \int_{r=0}^{\infty} f(r) \left[1 + \frac{\langle \eta^2 \rangle_{av}}{} \gamma(r) \right] \frac{\sin hr}{hr} r^2 \, dr \tag{40}$$

$$I_{V_v} = K_{10} \left\{ <\eta^2>_{av} \int_{r=0}^{\infty} \gamma(r) \frac{\sin hr}{hr} r^2 \, dr \right.$$

$$\left. + \frac{4}{45} <\delta^2>_{av} \int_{r=0}^{\infty} f(r) \left[1 + \frac{<\eta^2>_{av}}{\alpha^2} \gamma(r) \right] \frac{\sin hr}{hr} \right\} r^2 \, dr \quad (41)$$

where $<\delta^2>$ is the mean-squared anisotropy, $\overline{\alpha}$ is the average polarizability, and $f(r)$ is the orientation correlation function given by

$$f(r) = \frac{3 <\cos^2 \theta_{ij}>_r - 1}{2} \tag{42}$$

where θ_{ij} is the angle between the optic axes of the ith and j^{th}
scattering elements separated by a distance r.

Recently Yoon and Stein [29] have carried out extensive analysis and have determined expressions for scattered intensities from
an assembly of two- or three-dimensional spherulites using the cor-
relation-fluctuation approach. They started with Eq. (34) and obtained the vector form of Debye-Bueche equation for scattering
intensity:

$$I = K_{11} D <\eta^2> \int_0^{\infty} \gamma(\vec{r}) \exp[i \, k (\vec{r} \cdot \vec{s})] \, d\vec{r} \tag{43}$$

where D is the spherulite area A in two dimensions and spherulite
volume V in three dimensions. $\gamma(r)$ is the vector correlation function and is defined as before:

$$\gamma(\vec{r}) = \frac{<\eta_i \, \eta_j>_{\vec{r}}}{<\eta^2>} \tag{44}$$

For a system with spherical symmetry $\gamma(\vec{r}) = \gamma(r)$, and Eq. (43)
changes to

$$I = 4\pi K_{11} D <\eta^2> \int_0^{\infty} \gamma(r) \frac{\sin hr}{hr} r^2 \, dr \tag{45}$$

which is identical with Debye-Bueche equation with $K_9 = 4\pi K_{11} D$. Now
for an assembly of spherulites $<\eta^2>$ can be expressed as

$$\langle \eta^2 \rangle = \phi_s \langle \eta^2 \rangle_s + (1 - \phi_s) \langle \eta^2 \rangle_m \tag{46}$$

where ϕ_s is the fractional area (or volume) occupied by spherulites, $\langle \eta^2 \rangle_s$ and $\langle \eta^2 \rangle_m$ are the values of $\langle \eta^2 \rangle$ within and outside the spherulites, respectively. The value of $\eta(\vec{r}) = \rho(r) - \rho_0$ in turn is found from Eqs. (35), (38), (39), and (7) or (8) for H_v or V_v polarization, respectively.

A. Two-Dimensional Case

Considering a two-dimensional spherulite (anisotropic disk) with its plane perpendicular to the incident beam (Fig. 3), the unit vector $\vec{a}$ becomes

$$a = \sin \alpha \; \vec{j} + \cos \alpha \; \vec{k} \tag{47}$$

α being the angle between the polarization direction of the incident beam and the optic axis. For H_v polarization $(\vec{O}_{H_v} = - \sin \rho_2 \; \vec{i} + \cos \rho_2 \; \vec{j})$ using Eqs. (7), (38), (39), and (47) one obtains

$$[\rho(\vec{r})]_{H_v} = \frac{1}{2} C E_0 (\alpha_1 - \alpha_2) \cos \rho_2 \sin 2 \alpha \tag{48}$$

In the case of assembly of spherulites, if the volume element is within a spherulite which has its optic axis along its radius, then $\alpha_1 - \alpha_2 = \alpha_r - \alpha_t$; but if the volume element is in the isotropic medium surrounding the spherulites, then $\alpha_1 - \alpha_2 = 0$. ρ_0 may be obtained by averaging $\rho(r)$ over all values of α which occur with equal probability, thus

$$\rho_0 = \frac{CE_0}{4\pi} (\alpha_1 - \alpha_2) \cos \rho_2 \int_0^{2\pi} \sin 2\alpha \; d\alpha = 0 \tag{49}$$

and so

$$\eta(r) = \rho(r) \tag{50}$$

Moreover, since $\rho(r) = 0$ outside the spherulite,

$$\langle \eta^2 \rangle_{H_v} = \phi_s \langle \eta^2 \rangle_s = \phi_s \langle \rho^2 \rangle_s \tag{51}$$

$$= \frac{1}{4} \phi_s C^2 Eo^2 (\alpha_r - \alpha_t)^2 \cos^2 \rho_2 \langle \sin^2 2\alpha \rangle_s$$

where ϕ_2 is the fractional area occupied by spherulites, and

$$\langle \sin^2 2\alpha \rangle_s = \frac{1}{2\pi} \int_0^{2\pi} \sin^2 2\alpha \ d\alpha = \frac{1}{2}$$

Thus

$$\langle \eta^2 \rangle_{H_V} = \frac{1}{8} \phi_s C^2 E_0^2 (\alpha_r - \alpha_t)^2 \cos^2 \rho_2 \tag{52}$$

From Eqs. (44), (48), (50), and (52), the correlation function is found to be

$$[\gamma(\vec{r})]_{H_V} = 2 \langle \sin \ 2\alpha_i \sin 2\alpha_j \rangle_r \tag{53}$$

The averaging should be done over all pairs of volume elements both within a spherulite and in different spherulites in the assembly. Such an evaluation would have to be done numerically and would involve excessive computing time. Therefore, it is assumed that the correlation function for an assembly of spherulites is identical with that of a single spherulite, as was suggested by Sturgill [30] in his calculations for isotropic spheres. The scattered intensity is now given by Eq. (43)

$$I_{H_V} = K_{11} A \langle \eta^2 \rangle_{H_V} \int_0^\infty \gamma(\vec{r}) \ \exp[ik(\vec{r} \cdot \vec{s})] \ d\vec{r} \tag{54}$$

Calculations show that results are similar to those from model calculations discussed earlier. Comparing Eqs. (17) and (54), it is found that

$$I_{H_V} = \frac{1}{4} \pi K_{12} A \cos^2 \rho_2 \phi_s (\alpha_r - \alpha_t)^2 R^2 (\frac{2}{w^2})^2$$

$$\times \ [2 - 2 \ J_0(w) - w J_1(w)]^2 \ \sin^2 2\mu \tag{55}$$

It is evident that I_{H_V} is a function of ϕ_s, $(\alpha_r - \alpha_t)$ and the spherulitic radius R. A monotonic increase in I_{H_V} is predicted as ϕ_s varies from 0 to 1.

Similarly for V_V polarization ($\vec{0}_{V_V} = -\sin \rho_1 \vec{i} + \cos \rho_1 \vec{k}$) using Eqs. (8), (38), (39), and (47) one finds:

$$[\rho(r)]_{V_v} = CE_0 \cos \rho_1 [(\alpha_1 - \alpha_2) \cos^2\alpha + \alpha_2] \tag{56}$$

$$\rho_0 = CE_0 \cos \rho_1 [\frac{\phi_s}{2} (\alpha_r - \alpha_t) + (1 - \phi_s) \alpha_m] \tag{57}$$

where α_m is the polarizability of the medium outside the spherulites. Now,

$$[\eta_s]_{V_v} = CE_0 \cos \rho_1 [(\alpha_r - \alpha_t) \cos^2\alpha + (\alpha_t - \alpha_d)] \tag{58}$$

where

$$\alpha_d = \phi_s \frac{\alpha_r + \alpha_t}{2}] + (1 - \phi_s) \alpha_m \tag{59}$$

and

$$[\eta_m]_{V_v} = CE_0 \cos \rho_1 \phi_s \left[\frac{\alpha_r + \alpha_t}{2} - \alpha_m \right] \tag{60}$$

Proceeding as in the H_v case, I_{V_v} is found to be

$$
\begin{aligned}
I_{V_v} = \pi K_{13} AR^2 \cos^2\rho_1 \, B_1(\phi_s) \, (\frac{2}{w})^2 \Big[& (\alpha_r - \alpha_d) \, [1 - J_0(w)] \\
& + (\alpha_t - \alpha_d) \, \{wJ_1(w) - [1 - J_0(w)]\} \\
& - (\alpha_r - \alpha_t) \cos^2\mu \, \{2[1 - J_0(w)] - wJ_1(w)\} \Big]^2
\end{aligned}
\tag{61}
$$

where $B_1(\phi_s)$ is a concentration dependent factor given as

$$B_1(\phi_s) = \frac{\phi_s(\frac{3}{8} - \frac{\phi_s}{4})(\alpha_r - \alpha_t)^2 + \phi_s(1 - \phi_s)(\alpha_t - \alpha_m)(\alpha_r - \alpha_m)}{(\alpha_r - \alpha_t)^2(\frac{3}{8} - \frac{\phi_s}{2} + \frac{\phi_s^2}{4}) + (1 - \phi_s)^2(\alpha_t - \alpha_m)(\alpha_r - \alpha_m)} \tag{62}$$

Unlike the H_v intensity, I_{V_v} is dependent upon average polarizability of the system α_d (related to α_m) in addition to ϕ_s, $(\alpha_r - \alpha_t)$ and R. On examining Eq. (59) it can be seen that finite V_v intensity

is predicted even if the scattering entities were isotropic spheres
with $\alpha_r - \alpha_t = 0$. For anisotropic systems, as ϕ_s increases from 0
to 1 the V_V intensity goes through a maximum. A detailed discussion of the changes in H_V and V_V scattering in crystallizing systems is presented later.

B. Three-Dimensional Case

As in the model calculations, a three-dimensional spherulite
is considered to be a homogeneous anisotropic sphere. Procedures
used for evaluating correlation functions and scattering intensities
are the same as outlined for two-dimensional spherulites. Sturgill's
approximation that the correlation function for an assembly of spherulites is identical with that of a single spherulite is again made.
For calculations in three dimensions the definitions of the unit
vector a and the average polarizability α_d are changed as

$$\vec{a} = (\sin \alpha \cos \Omega)\vec{i} + (\sin \alpha \sin \Omega)\vec{j} + (\cos \alpha)\ \vec{k} \qquad (63)$$

$$\alpha_d = \phi_s \frac{\alpha_r + 2\alpha_t)}{3} + (1 - \phi_s)\ \alpha_m \qquad (64)$$

It should be noted that ϕ_s is now the volume fraction occupied by
the spherulites. Proceeding as before, the scattering intensities
are found to be:

$$I_{H_V} = \frac{1}{4}\ \phi_s K_{14} V \cos^2 \rho_2 (\alpha_r - \alpha_t)^2 R^3 (\frac{3}{U^3})^2 \left\{ (\alpha_r - \alpha_t) \frac{\cos^2 (\frac{\theta}{2})}{\cos\theta} \right.$$
$$\left. \sin 2\mu\ [4 \sin U - U \cos U - 3\ \text{Si}\ U] \right\}^2 \qquad (65)$$

and

$$I_{V_V} = K_{15} V \pi R^3 \cdot B_2(\phi_s)\ \cos^2 \rho_1 (\frac{3}{U^3})^2 \left[(\alpha_t - \alpha_d)(2 \sin U \right.$$
$$- U \cos U - \text{Si}\ U) + (\alpha_r - \alpha_d)(\text{Si}\ U - \sin U)$$
$$+ (\alpha_r - \alpha_t)\frac{\cos^2 (\frac{\theta}{2})}{\cos\ \theta}$$
$$\left. \cos^2\mu\ (4 \sin U - U \cos U - 3\ \text{Si}\ U) \right]^2 \qquad (66)$$

where $B_2(\theta_s)$ is a concentration dependent factor given as

$$B_2(\phi_s) = \frac{\left\{\cos^2\rho_1[\phi_s(1-\phi_s)(\alpha_t - \alpha_m)(\tfrac{2}{3}\alpha_r + \tfrac{1}{3}\alpha_t - \alpha_m) + \phi_s(\tfrac{1}{5} - \tfrac{\phi_s}{9})(\alpha_r - \alpha_t)^2] + \sin^2\rho_1\tfrac{\phi_s}{15}(\alpha_r - \alpha_t)^2\right\}}{\left\{\cos^2\rho_1[(1-\phi_s)^2(\alpha_t - \alpha_m)(\tfrac{2}{3}\alpha_r + \tfrac{1}{3}\alpha_t - \alpha_m) + (\alpha_r - \alpha_t)^2(\tfrac{1}{5} - \tfrac{2\phi_s}{9} + \tfrac{\phi_s^2}{9})] + \sin^2\rho_1\tfrac{\phi_s}{15}(\alpha_r - \alpha_t)^2\right\}}$$

(67)

Similar to the two-dimensional results, the I_{H_V} is a function of ϕ_s, $(\alpha_r - \alpha_t)$, R, and α_d (or α_m).

The evaluation of H_V and V_V intensities can be suitably applied to characterize the crystallization process in polymeric systems. During the course of crystallization, the H_V intensity will increase monotonically because of an increase in the spherulitic radius R, the volume fraction of spherulites ϕ_s, and the anisotropy $(\alpha_r - \alpha_t)$ of the spherulite. Any increase in intensity after ϕ_s has attained a value of unity would be due to an increase in the anisotropy of the spherulite $(\alpha_r - \alpha_t)$ during secondary crystallization [31]. As an example, Fig. 12 illustrates the changes occurring in the H_V patterns obtained during the crystallization of polyethylene terephthalate at 110°C.

On the other hand, the V_V scattering depends upon the function $B(\phi_s)$ [$B_1(\phi_s)$ in two dimensions and $B_2(\phi_s)$ in three dimensions] which makes the intensity pass through a maximum as ϕ_s varies from 0 to 1, as long as the polarizability of the surrounding medium is not between the radial and tangential polarizabilities of the spherulite. In other words, either $\alpha_m < \alpha_r$, α_t; or $\alpha_m > \alpha_r$, α_t which is usually true for crystalline polymers. The maxima in V_V intensity would generally occur when the system is about half filled with spherulites. However, the exact value at the maxima clearly depends upon the values of α_r, α_t, and α_m. It should be noted that during the course of crystallization, increase in spherulitic radius R, as

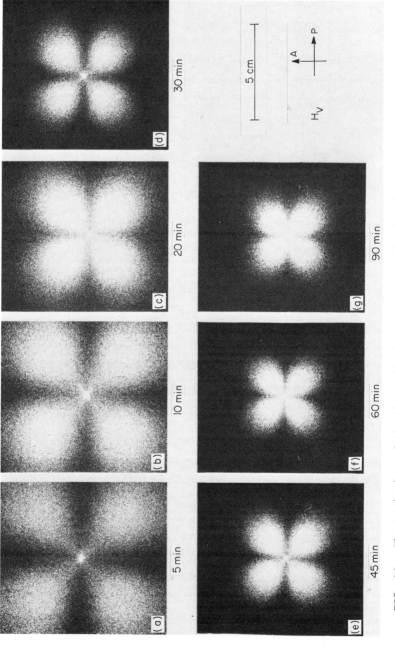

FIG. 12. The variation of the H_V-scattering patterns with time during the isothermal crystallization of polyethylene terephthalate at 110°C.

well as the anisotropy $(\alpha_r - \alpha_t)$, also contributes to the V_v scattering and are entirely responsible for the intensity when ϕ_s is unity. As ϕ_s attains the value of unity, $B(\phi_s) = \phi_s$ and the contribution of surrounding polarizability diminishes. At this point the V_v intensity goes through a minimum and then rises a second time due to a further increase in anisotropy $(\alpha_r - \alpha_t)$ during secondary crystallization [31]. An illustration of the variation of V_v intensities for polyethylene terephthalate crystallizing at 110°C is shown in Fig. 13.

The theory extends further to predict the shape of V_v scattering patterns during crystallization for two-dimensional and three-dimensional spherulites. In the two-dimensional case, when the volume fraction of spherulites ϕ_s is small the scattering arises from density fluctuation and a circular pattern is predicted. As ϕ_s increases, the pattern develops a twofold symmetry and is elongated along the polarization direction. When ϕ_s approaches unity, i.e., the spherulites become volume filling, the twofold symmetry changes to a fourfold symmetry. On the other hand, for the three-dimensional case, the circular pattern is predicted as before during the early stages when ϕ_s is small. As ϕ_s increases and the spherulites become volume filling, the pattern changes, takes on a twofold symmetry being elongated in the polarization direction, and maintains this twofold symmetry. The actual shape of the pattern depends upon the anisotropy of the spherulite and the polarizability difference between the spherulite and its amorphous surroundings [6,29,32].

IV. SCATTERING FROM RODS

A. Model Approach

The model approach has been applied to a rodlike geometry in order to account for the scattering from some polymeric systems which do not exhibit a spherulitic morphology. Stein and Rhodes [33] derived the basic theory for scattering from two-dimensional anisotropic rods. The rods are considered to be of infinitismal

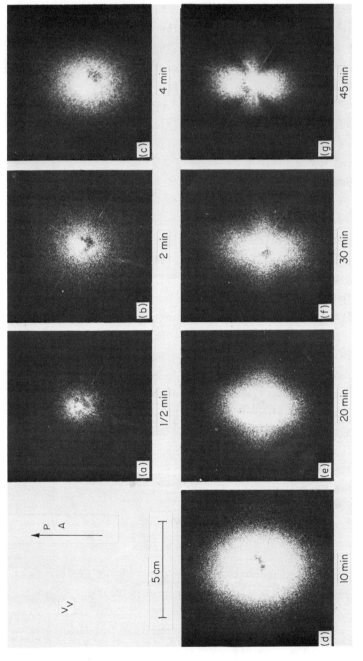

FIG. 13. The variation of the V_v-scattering patterns with time during the isothermal crystallization of polyethylene terephthalate at 110°C.

thickness and length L lying in a plane perpendicular to the incident beam and tilted at an angle α with respect to the vertical direction (Fig. 14). The scattering amplitude from a differential
area element of this single rod is

$$dE(\alpha) = \rho \exp[ik(\vec{r} \cdot \vec{s})] \, dr \tag{68}$$

where ρ is the scattering power of the rod, k, $\vec{r}$, and $\vec{s}$ are the
same as defined in the spherulitic case. By integrating $dE(\alpha)$ over
all volume elements (from -L/2 to L/2 one gets

$$E(\alpha) = \rho L \frac{\sin(kaL/2)}{kaL/2} \tag{69}$$

where

$$a = \sin\theta \, \sin(\alpha + \Omega) \tag{70}$$

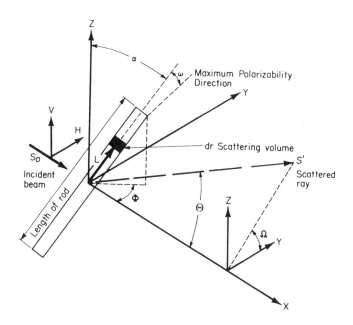

FIG. 14. Coordinate system for rod scattering [from M. B.
Rhodes and R. S. Stein, *J. Polymer Sci.*, A-2, *1*, 1539 (1969), Fig.6].

The scattering power for H_v and V_v polarizations are given as

$$\rho_{H_v} = \rho_0 (\alpha_1 - \alpha_2) \sin(\alpha + \omega) \cos(\alpha + \omega) \tag{71}$$

$$\rho_{V_v} = \rho_0 [(\alpha_1 - \alpha_2) \cos^2(\alpha + \omega) + \alpha_2] \tag{72}$$

where ρ_0 is the scattering power per unit length per unit incident field strength, α_1 and α_2 are the polarizabilities along and perpendicular to the optic axis, and ω is the angle between the optic axis and the rod axis. For an assembly of rods scattering independently, the scattering intensity is now given as

$$I = \int N(\alpha) \, E^2 \, d\alpha \tag{73}$$

where $N(\alpha)$ is a distribution which may be assumed to be [33]

$$N(\alpha) = N_0 (\varepsilon^{-2} \cos^2 \alpha + \varepsilon^2 \sin^2 \alpha)^{-1/2} \tag{74}$$

where ε is a parameter characterizing the degree of orientation. For random orientation $\varepsilon = 1$ and $N(\alpha) = N_0$. Some calculated H_v patterns with different values of ω are presented in Fig. 15. It is seen that, unlike the spherulitic case, there is no maxima in scattering intensity. The intensity has its maximum value at the center and decreases monotonically with scattering angle. Similar to the spherulitic case, the pattern has four lobes which are oriented along the polar directions for $\omega = 45°$ and are at $45°$ to polars for ω close to $0°$ or $90°$. Such patterns are experimentally observed in polytetrafluoroethylene (teflon) [33] and collagen [34, 35]. Kawai et al. [35,36] have extended the model to three dimensions and have noticed little difference in the results.

Recently Prud'homme and Stein [37] have developed a theory to take into account the interference effect among rods using the correlation-function approach.

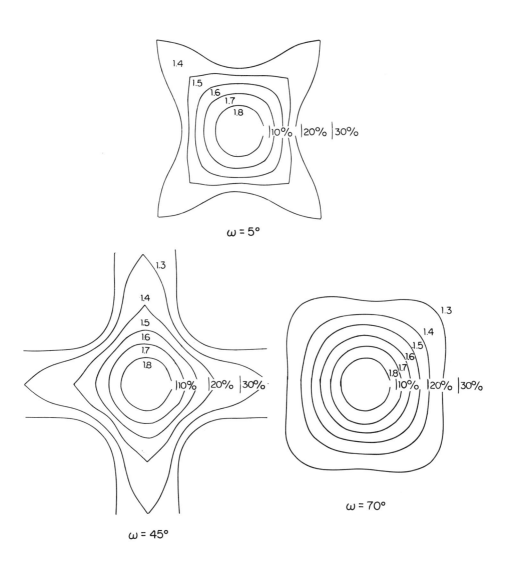

FIG. 15.   Calculated H_V-scattering patterns for anisotropic rods for
(a) $\omega = 5°$, (b) $\omega = 45°$, and (c) $\omega = 70°$ [from M. B. Rhodes and R. S.
Stein, *J. Polymer Sci.*, A-2, *1*, 1539 (1969), Fig. 6].

B. Oriented Systems

The stretching of spherulitic polymers results in the deformation of spherulites, and it can be assumed that the circular spherulite transforms into an ellipsoid. Clough et al. [8] have proposed a two-dimensional theory for scattering from ellipsoids to
account for experimentally observed patterns from oriented polyethylene [4]. Their theory has been extended to a generalized
three-dimensional model by van Aartsen and Stein [38]. Samuels
[39] has proposed a relatively simple semi-empirical theory to
account for the change in scattering patterns upon deformation.
His results compare favorably with patterns obtained from oriented
polypropylene, as can be seen in Fig. 16.

REFERENCES

1. R. S. Stein, in "Proceeding of the Interdisciplinary Conference on Electromagnetic Scattering," (M. Kerker, ed.) Pergamon
   Press, New York, 1963, pp. 430-458.

2. R. S. Stein, P. Erhardt, S. Clough, and J. J. van Aartsen, in
   "Electromagnetic Scattering," (R. L. Rowell and R. S. Stein,
   eds.), Gordon and Breach, New York, 1967, pp. 339-410.

3. R. S. Stein, in "Structure and Properties of Polymer Films,"
   (R. W. Lenz and R. S. Stein, eds.), Plenum Press, New York,
   1973, pp. 1-24.

4. R. S. Stein and M. B. Rhodes, *J. Appl. Phys.*, *31*, 1873 (1960).

5. R. J. Samuels, *J. Polymer Sci.*, A-2, *9*, 2165 (1971).

6. R. S. Stein, S. N. Stidham, and P. R. Wilson, ONR Technical
   Report No. 36, Project: 356-378, Contract No. 3557(01),
   University of Massachusetts, Amherst, Massachusetts (1961).

7. R. S. Stein, M. B. Rhodes, P. R. Wilson, and S. N. Stidham,
   Pure Appl. Chem., *4*, 219 (1962).

8. S. Clough, J. J. van Aartsen, and R. S. Stein, *J. Appl. Phys.*,
   36, 3072 (1965).

9. R. S. Stein and C. Picot, *J. Polymer Sci.*, A-2, *8*, 1955 (1970).

10. R. E. Prud'homme and R. S. Stein, *J. Polymer Sci.*, A-2, *11*,
    1357 (1973).

11. C. Picot, R. S. Stein, R. M. Marchessault, J. Borch, and A.
    Sarko, *Macromolecules*, *4*, 467 (1971).

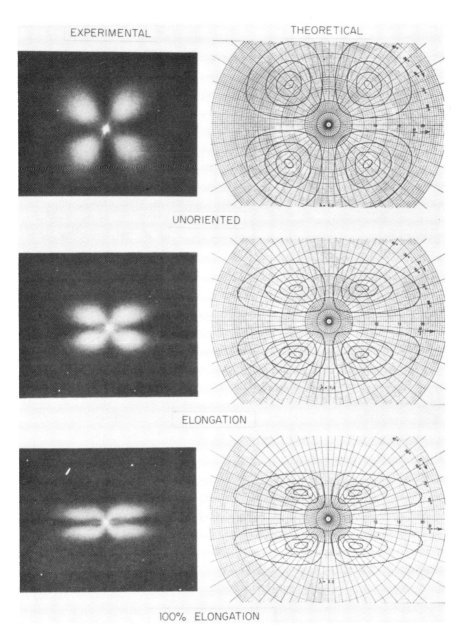

FIG. 16. Experimental and theoretically predicted H_V patterns from deformed polypropylene spherulites. The stretching direction is vertical [from R. J. Samuels, *J. Polymer Sci.*, *C13*, 37 (1966), Figs. 2-39].

12. R. S. Stein and C. Picot, *J. Polymer Sci.*, A-2, *8*, 2127 (1970).

13. R. E. Prud'homme and R. S. Stein, *J. Polymer Sci.*, A-2, *11*, 1683 (1973).

14. M. Motegi, T. Oda, M. Moritani, and H. Kawai, *Polymer Journal*, (Japan), *1*, 209 (1970).

15. C. Picot, R. S. Stein, M. Motegi, and H. Kawai, *J. Polymer Sci.*, A-2, *8*; 2115 (1970).

16. A. Misra and R. S. Stein, *J. Polymer Sci.*, A-2, *11*, 109 (1973).

17. R. E. Prud'homme, D. Y. Yoon, and R. S. Stein, *J. Polymer Sci.*, A-2, *11*, 1047 (1973).

18. R. S. Stein, P. R. Wilson, and S. N. Stidham, *J. Appl. Phys.*, *34*, 46 (1963).

19. A. E. M. Keijzers, J. J. van Aartsen, and W. Prins, *J. Amer. Chem. Soc.*, *90*, 3107 (1968).

20. R. S. Stein and W. Chu, *J. Polymer Sci.*, A-2, *8*, 1137 (1970).

21. T. Hashimoto and R. S. Stein, *J. Polymer Sci.*, A-2, *9*, 1747 (1971).

22. D. Y. Yoon and R. S. Stein, *J. Polymer Sci.*, A-2, *12*, 763 (1974).

23. R. S. Stein and J. J. Keane, *J. Polymer Sci.*, *17*, 21 (1955).

24. R. E. Prud'homme, L. Bourland, R. T. Natarajan, and R. S. Stein, *J. Polymer Sci.*, A-2, *12*, 1955 (1974).

25. R. E. Prud'homme, R. T. Natarajan, L. Bourland, and R. S. Stein, *J. Polymer Sci.*, *Polym. Phys. Ed.*, *14*, 1541 (1970).

26. P. Debye and A. Bueche, *J. Appl. Phys.*, *26*, 518 (1949).

27. M. Goldstein and E. R. Michalik, *J. Appl. Phys.*, *26*, 1450 (1955).

28. R. S. Stein and P. R. Wilson, *J. Appl. Phys.*, *33*, 1914 (1962).

29. D. Y. Yoon and R. S. Stein, *J. Polymer Sci.*, A-2, *12*, 733 (1974).

30. D. T. Sturgill, paper presented at the American Ceramic Society Symposium on Nucleation, 1971; "Advances in Nucleation and crystallization in Glasses," American Ceramic Society, Special Publ. No. 5, (L. L. Hench and S. W. Freidman, eds.), 1972.

31. L. Mandelkern, "Crystallization of Polymers," McGraw-Hill, New York, 1964.

32. R. J. Samuels, *J. Polymer Sci.*, A-2, *12*, 1417 (1974).

33. M. B. Rhodes and R. S. Stein, *J. Polymer Sci.*, A-2, *1*, 1539 (1969).

34. J. C. W. Chien and E. P. Chang, *Macromolecules*, *5*, 610 (1972).

35. M. Moritani, N. Hayashi, A. Utsuo, and H. Kawai, *Polymer J.*, *2*, 74 (1971).

36. N. Hayashi and H. Kawai, *Polymer J.*, *3*, 140 (1972).

37. R. E. Prud'homme and R. S. Stein, *J. Polym. Sci., Polym. Phys. Ed.*, *12*, 1805 (1974).

38. J. J. van Aartsen and R. S. Stein, *J. Polymer Sci.*, A2, *9*, 295 (1971).

39. R. J. Samuels, *J. Polymer Sci.*, *C13*, 37 (1966).

AUTHOR INDEX

Numbers in parentheses are reference numbers and indicate that an author's work is referred to although his name is not cited in text. Underlined numbers give the page on which the complete reference is listed.

SUBJECT INDEX

Acrylonitrile, 127, 167
Adsorption, 5
Amplitude, scattering, 325
Amylopectin, degradation, 108
Analog computer, 236, 260, 308
Analog-to-digital converter, 207, 239
Anisotropy, 324, 338, 330
Assembler, 237
Autoaccelerating, 8, 10, 18, 20, 22
"Azeotropic" polymers, 77

Bernoullian statistics, 8, 34, 37, 42, 45, 46, 50, 62
Bessel functions, 331
Binary code, 237
Biopolymers, 103
Bit, 237
Block character, 8, 10, 28, 43, 45, 50
Block length, average, 3, 51
Butadiene, 127

CDC 6600 terminal, 235
Central processor, 206
Chain branching, 104
Charge transfer, 158, 160
p-Chlorostyrene, 182
Compatibility testing, 283
Compiler, 237
Composition distribution, 103, 155, 158
Computation time, 4, 18, 25, 80, 122, 151, 236, 313, 341

Conditional probability, 40, 60, 62, 64, 69
  matrix of, 70, 71, 81
Coniferyl alcohol, 106
Contract charge transfer, 165
Conversational Language for Spectroscopic System (CLASS), 255
Conversion, 7, 8, 25, 37, 77, 146, 153
Copolymerization, 66, 100, 118, 119, 146
  multicomponent, 118
  reversible, 77, 103
  terminal model, 40, 147, 154, 161
Copolymers, 13
Cost of simulation, 109
β-cyanoacrolein, 164

Decomposition, first order kinetics of teflon, 281
Degradation, 107
Degree of polymerization, 10
2,5-Dichlorostyrene, 181, 184
Differential method for DSC and TGA analysis, 276
Digital minicomputers, 236
Digital-to-analog converter, 207, 239
1,1-Diphenylpicrylhydrazyl, 288
Dipole moment, induced, 327, 345
Disks, 324, 330
Disorder effect, 325, 338
DSC, 251, 273